Joe E. Coleman

A HANDBOOK ON ELECTROMAGNETIC SHIELDING MATERIALS AND PERFORMANCE

By Donald R.J. White, MSEE/PE

Interference Control Technologies, Inc.
Don White Consultants, Subsidiary
State Route 625, P.O. Box D
Gainesville, Virginia 22065 USA
Telephone (703) 347-0030 Telex 89-9165 DWCI GAIV

Second Edition
Fifth Printing 1988

Library of Congress Catalog Card No. 75-16592
ISBN# 0-932263-08-9

Printed in the United States of America

ACKNOWLEDGEMENT

The author wishes to thank the many people who encouraged him to write this handbook on shielding. He expresses his appreciation to the individuals and companies who have furnished several of the illustrative figures, for which acknowledgements in this handbook have been made.

The author expresses his appreciation to his wife Colleen and Muriel M. Moeller for their assistance in typing, and in the many facets of logistics involved in preparation of the manuscript; to Luis F. Longoria III and Jane Backstrom for their drafting, HP-65 calculator computations, and the many others who have helped produce this publication.

Handbooks Published by DWCI

(1) White, Donald R.J., *Electrical Filters—Synthesis, Design & Applications,* 1980.

(2) White, Donald R.J., Volume 1, *Electrical Noise and EMI Specifications,* 1971.

(3) White, Donald R.J., Volume 2, *Electromagnetic Interference Test Methods and Procedures,* 1980.

(4) White, Donald R.J., Volume 3, *Electromagnetic Interference Control Methods & Techniques,* 1973.

(5) White, Donald R.J., Volume 4, *Electromagnetic Interference Test Instrumentation Systems,* 1980.

(6) Duff, Dr. William G. and White, Donald R.J., Volume 5, *Electromagnetic Interference Prediction & Analysis Techniques,* 1972.

(7) Hill, James S. and White, Donald R.J., Volume 6, *Electromagnetic Interference Specifications Standards & Regulations,* 1975.

(8) White, Donald R.J., *A Handbook on Electromagnetic Shielding Materials and Performance,* 1980.

(9) Duff, Dr. William G., *A Handbook on Mobile Communications,* 1980.

(10) White, Donald R.J., *EMI Control Methodology & Procedures,* 1982.

(11) White, Donald R.J., *EMI Control in the Design of Printed Circuit Boards and Backplanes,* 1982.

(12) Jansky, Donald M., *Spectrum Management Techniques,* 1977. (Also available in French language.)

(13) Herman, John R., *Electromagnetic Ambients and Man-Made Noise,* 1979.

(14) Hart, William C. and Malone, Edgar W., *Lightning and Lightning Protection,* 1979.

(15) Kaiser, Dr. Bernhard E., *EMI Control in Aerospace Systems,* 1979.

(16) Feher, Dr. Kamilo, *Digital Modulation Techniques in an Interference Environment,* 1977.

(17) Gard, Michael F., *Electromagnetic Interference Control in Medical Electronics,* 1979.

(18) Carstensen, Russell V., *EMI Control in Boats and Ships,* 1979.

(19) Georgopoulos, Dr. Chris J., *Fiber Optics and Optical Isolators,* 1982.

(20) Mardiguian, Michel, *How to Control Electrical Noise,* 1983.

(21) Denny, Hugh W., *Grounding for Control of EMI,* 1983.

(22) Ghose, Rabindra N., *EMP Environment and System Hardness Design,* 1983.

(23) Mardiguian, Michel, *Interference Control in Computers and Microprocessor-Based Equipment,* 1984.

(24) *EMC Technology 1982 Anthology*

Notice

All of the books listed above are available for purchase from Don White Consultants, Inc., State Route 625, P.O. Box D, Gainesville, Virginia 22065 U.S.A. Telephone: (703) 347-0030; Telex:89-9165 DWCI GAIV.

PREFACE

There exists substantial material in the literature on the subject of electromagnetic shielding. Chap. 4 presents many references. For either an individual who has only recently been introduced to shielding or to a design engineer, however, much of the literature appears to be either confusing or poorly organized for design use. Missing in the literature are a series of useful design graphs including all the principal variables presented in a clear, understandable manner. Thus, this handbook on Shielding was conceived to fill these voids.

This handbook does not cover the topics of where and when to shield and where to ground a shield. These topics are covered in Vol. 3 of the EMC Handbook Series. Rather, this handbook explains shielding theory and performance and presents many design graphs of shielding effectiveness vs frequency as a function of shield metal and its characteristics, and E and H-fields and plane waves.

Regarding the impedance of the fields (E, H, or plane waves), the literature and manufacturers' data are often very misleading. For example, since the wave and circuit impedance which produced the field are interlocked and since a circuit impedance is not infinite, E-field shielding effectiveness data are generally optimistic (too high) relative to actual performance. In a converse manner, H-field shielding effectiveness data are pessimistic (too low) since a magnetic source circuit impedance is not zero. This handbook clarifies and quantifies these points.

Another example of possibly misleading information is the use of MIL-STD-285 to measure and report the shielding effectiveness of test items to E and H-fields. The reference test distance per MIL-STD-285 is one foot (0.305 meters). Thus, for installations located in the near field which are greater than one foot from an interfering source, actual E-field shielding performance will be less and H-field performance will be greater than that reported by MIL-STD-285 measurements. The converse applies for application distances between sources and metal barriers which are less than one foot away as illustrated in this handbook.

The discussions and design data on shielding effectiveness in this handbook are not restricted to homogeneous metals. In fact no real life and useful shielded compartment, box, cabinet, or room is homogeneous since usually many penetrations of a six-sided shield configuration are necessary. Techniques used to reinstate the integrity of a shielded enclosure are discussed in Vol. 3 of the EMC Handbook Series. Shielding materials and performance of non-homogeneous metals are discussed in this handbook on Shielding. Some examples are pseudo-homogeneous shields made from metal deposition and flame-spray processes. Shields made of small-aperture metals are also presented. Examples include screens,

PREFACE

wire meshes, cable braids and metalized textiles, all of which are discussed herein together with design data.

The appendices of this handbook are perhaps the most important of all material presented. They contain 42 pages of design shielding effectiveness graphs for several metals whose thicknesses range from 0.0001 mil (2.54 nm) to 1 inch (2.54 cm). For both near and far-field calculations and associated frequencies, the design graphs cover source-to-metal distances ranging from 10 cm to 10 km. Frequency coverage is from 10 Hz to 30 GHz. All data were run-off on the HP-65 programmable calculator. For those who have an HP-65, the program is presented so that they can develop and use their own magnetic card. There also exist many design graphs other than direct shielding effectiveness which the reader should find useful.

The author of this handbook invites the user to communicate with him. He especially invites comments, questions, or requests for further elucidation.

December 1975 — Donald R. J. White
Germantown, Maryland USA

January 1980 — Second Edition
Gainesville, Virginia USA

TABLE OF CONTENTS

ELECTROMAGNETIC SHIELDING MATERIALS AND PERFORMANCE

LIST OF TABLES

LIST OF ILLUSTRATIONS

CHAPTER 1 SHIELDING THEORY

CHAPTER 2 SHIELDING MATERIALS AND TESTING

List of Illustrations

APPENDIX A COPPER

APPENDIX B MONEL

List of Illustrations

APPENDIX F 78 PERMALLOY

APPENDIX G HIGH PERMEABILITY

LIST OF SYMBOLS AND ABBREVIATIONS

A_{dB} = absorption loss in dB

B_{dB} = re-reflection loss in dB

c = velocity of electromagnetic wave in air = $1/\sqrt{\mu\varepsilon}$ = 3×10^8 m/sec.

cm = centimeter = 0.01 meter = 0.3937 inches

Cu = copper

dB = decibel = 0.1 Bel = 10 $\log_{10}$(power ratio)

e = Napierian base = 2.718

E = electric-field strength in volts/meter

f = frequency in Hertz

Fe = iron

f_{MHz} = frequency in MHz

H = magnetic-field strength in amperes/meter

I = current in amperes

j = imaginary operator = angle $\pi/2$ = 90 degrees

k = $\lambda/2\pi r$ for E fields; $2\pi r/\lambda$ for H fields; = 1 for plane waves

K = wave-to-metal impedance ratio, Z_w/Z_m = VSWR for $K \geq 1$

m = meter = 100 cm = 1000 mm = 39.37 inches = 3.28 feet

mil = 0.001 inch = 2.54×10^{-3} cm = 25.4 μm

mm = millimeter = 0.1 cm = 0.001 meter = 39.37 mils

nm = nanometer = 10^{-9} m = 10^{-6} mm = 10^{-3} μm = 39.37×10^{-6} mils

r = distance from EMI emission source to metal barrier

r_m = distance r in meters

SE_{dB} = shielding effectiveness (loss) in dB

R_{dB} = reflection loss in dB (excludes re-reflection loss)

t = metal thickness

t = time in seconds

t/δ = ratio of metal-thickness to skin-depth

V = voltage in volts

VSWR = voltage standing wave ratio

Z = impedance in ohms

Z_b = barrier metal impedance, Z_m for any t/δ ratio

Z_c = circuit impedance in ohms

List of Symbols and Abbreviations

Z_m = impedance of metal for $t/\delta >> 1$

Z_o = plane-wave impedance = $\sqrt{\mu_o/\varepsilon_o}$ = 120π = 377 ohms

Z_w = E/H = wave impedance

α = attenuation constant

β = phase constant

γ = propagation constant = $\alpha + j\beta$

Γ = voltage or wave transmission coefficient

Γ_{am} = transmission coefficient from air to metal interface

Γ_{ma} = transmission coefficient from metal to air interface

δ = skin depth in metal

ε = permitivity of medium = $\varepsilon_o \varepsilon_r$

ε_o = absolute permitivity of air = $1/36\pi x 10^9$ farads/m

ε_r = permitivity relative to air

λ = wavelength = c/f

μ = permeability of medium = $\mu_o \mu_r$

μ_o = absolute permeability of air = $4\pi x\ 10^{-7}$ henrys/m

μ_r = permeability relative to air (or copper)

μ_m = micrometer = 10^{-6}m = 10^{-3}mm = 0.3937 mils

ρ = voltage or wave reflection coefficient

ρ_{am} = reflection coefficient from air to metal interface

ρ_{ma} = reflection coefficient from metal to air interface

σ = conductivity of medium in mhos per unit distance

σ_r = conductivity relative to copper

ω = radial frequency in radians/sec = $2\pi f$

Ω = impedance in ohms

Ω/sq = surface impedance (or resistance) in ohms/square

CHAPTER 1

SHIELDING THEORY

Eighty per cent of this handbook is design data of shielding effectiveness vs. several variables including metal type, metal properties, thickness, distance, frequency, etc. Most of the design and applications data are presented in the appendices.

Basics and fundamentals of shielding are presented in Chaps. 1 and 2. Chap. 3 covers information on how to use the design graphs in the appendices including constraints and illustrative examples. This chapter, Chap. 1, presents tutorial information on shielding theory and materials. This includes field theory, near and far-field definitions, wave impedance, metal barrier impedance, absorption loss, reflection loss, and overall shielding effectiveness.

1.1 FIELD THEORY

The purpose of this section is to present some relations about magnetic, electric, and electromagnetic fields as pertinent background to understanding and applying field theory. Since the literature is replete with discussions of Maxwell's equations and field theory, only a few aspects are presented here.

The electric (E_θ, E_r) and magnetic (H_ϕ) fields existing about an oscillating doublet (tiny dipole in which its length ($D \ll \lambda$) are obtained from applying Maxwell's equations:

$$E_\theta = \frac{Z_0 ID\pi \sin\theta}{\lambda^2}\left[-\left(\frac{\lambda}{2\pi r}\right)^3 \cos\psi - \left(\frac{\lambda}{2\pi r}\right)^2 \sin\psi + \left(\frac{\lambda}{2\pi r}\right)\cos\psi\right] \qquad (1.1)$$

$$E_r = \frac{2Z_0 ID\pi \cos\theta}{\lambda^2}\left[\left(\frac{\lambda}{2\pi r}\right)^3 \cos\psi + \left(\frac{\lambda}{2\pi r}\right)^2 \sin\psi\right] \qquad (1.2)$$

$$H_\phi = \frac{ID\pi \sin\theta}{\lambda^2}\left[\left(\frac{\lambda}{2\pi r}\right)^2 \sin\psi + \left(\frac{\lambda}{2\pi r}\right)\cos\psi\right] \qquad (1.3)$$

where, Z_o = free-space impedance (for $r \gg \lambda/2\pi = \sqrt{\mu/\varepsilon} = 120\pi = 377\Omega$)

I = current in short wire (doublet)
D = length of short wire (doublet) in which $D \ll \lambda$
θ = zenith angle to radial distance r
λ = wavelength corresponding to frequency, $f = c/\lambda$
r = distance from short wire doublet to measuring or observation point
$\psi = 2\pi r/\lambda - \omega t$
ω = radial frequency = $2\pi f$
t = time = 1/f
$c = 1 \sqrt{\mu\varepsilon} = 3x10^8 m/sec$

Several observations can be made about the near and far fields from Eqs. (1.1) and (1.3):

(1) When the multiplier, $\lambda/2\pi r = 1$ in the electric-field and the magnetic-field terms, all coefficients of either sine or cosine terms are unity and equal. Thus, when $r = \lambda/2\pi$ (about one-sixth wavelength), this corresponds to the transition-field condition or boundary between the near field (first term of equations) and far field (last term).

(2) When $r \gg \lambda/2\pi$ (far-field conditions), only the last term of Eqs. (1.1) and (1.3) is significant*. For this condition the wave impedance, $Z_o = E_\theta/H_\phi = 377$ ohms. This is called the radiation field (plane waves) and both E_θ and H_ϕ are in time phase ($\cos\psi$) although they are in directional quadrature.

(3) When $r \ll \lambda/2\pi$ (near-field conditions), only the first term of each equation is significant*. For this condition the wave impedance $E_\theta/H_\phi = Z_o\lambda/2\pi r$. Note that the wave impedance is now $\gg Z_o$ since $\lambda/2\pi r \gg 1$. This is sometimes called an electric field or a *high-impedance field*, i.e., high relative to a plane-wave impedance. It is also the induction field and E_θ and H_ϕ are in quadrature of both time phase and direction.

(4) Had the oscillating source not been a small straight wire or doublet exhibiting high-circuit impedance, but rather a small wire loop exhibiting low-circuit impedance, the first term appearing in Eqs. (1.1) and (1.2) would vanish, and a similar first term would appear in Eq. (1.3). For this condition, the wave impedance in the near field, $E_\theta/H_\phi = Z_o 2\pi r/\lambda$. Note that the wave impedance is now $\ll Z_o$ for $\lambda/2\pi r \gg 1$. This is sometimes called a magnetic field or a *low-impedance field*, i.e., low relative to Z_o, the plane-wave (radiation) impedance.

Fig. 1.1 illustrates the above first three observations (#1, 2, and 3) for the amplitude of each of the electric-field terms in Eq. (1.1). Note that the quasi-stationary field is the largest term in the near field and the induction term is next largest, whereas, the

* E_r is dropped from further discussion here.

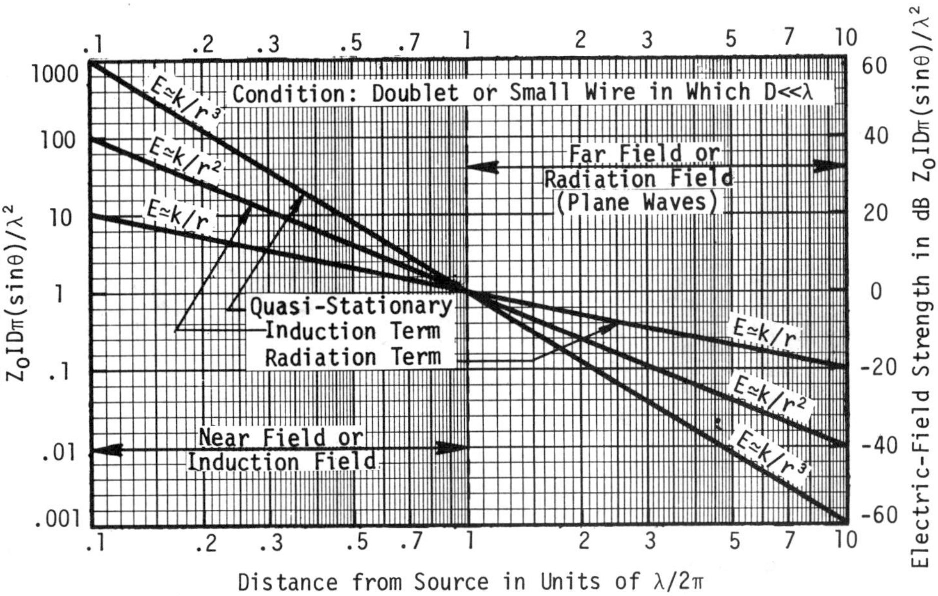

Figure 1.1 - Electric-Field Strength vs. Source Distance

radiation term is the largest in the far field. For $r = \lambda/2\pi$, coefficients of all three terms are equal.

Fig. 1.2 illustrates the preceding 3rd and 4th observations in the near or induction field. Situation (a) in Fig. 1.2 is a short monopole or straight wire in which the circuit current is low*. Consequently, the source or circuit impedance = V/I is a high impedance. As previously shown, the wave impedance near-in is also high, relative to 377 ohms. The associated electric field attenuates more rapidly ($1/r^3$) with an increase in distance than the magnetic field ($1/r^2$) in the induction region (cf. Eqs. (1.1) and (1.3). Thus the wave impedance decreases linearly (20 dB per decade) with an increase in distance where it asymptotically approaches Z_o=377 ohms in the far or radiation field for $r \gg \lambda/2\pi$. This relation is depicted in Fig. 1.3.

The converse to the above applies for situation (b) in Fig. 1.2 wherein a low-impedance source or circuit creates a low-impedance wave, i.e., a magnetic field. This wave impedance increases linearly (20 dB per decade) with distance where it asymptotically approaches 377 ohms

---

* The conduction current in the wire is the same as the displacement current flowing in the capacitance from wire to ground or from one end of the wire to the lower return side.

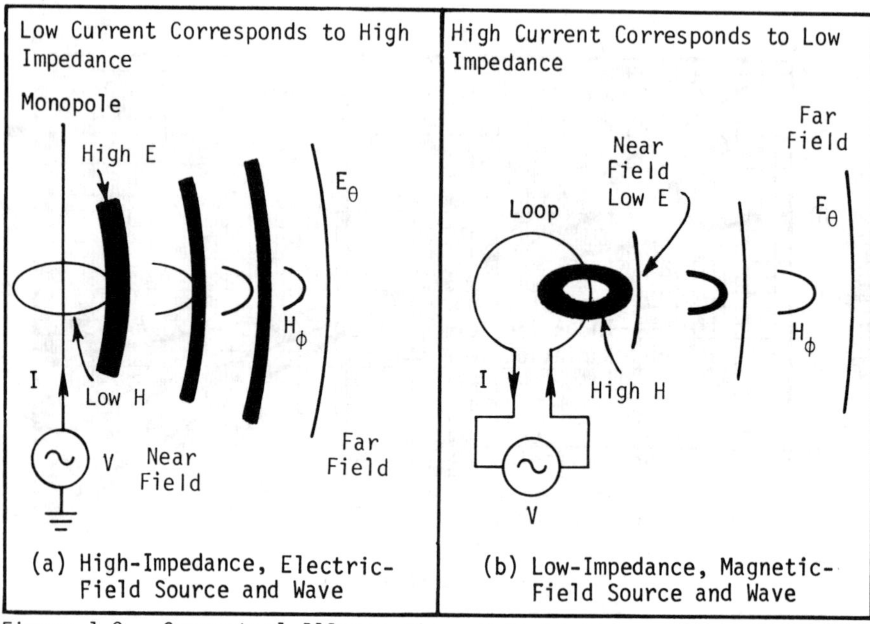

Figure 1.2 - Conceptual Illustration of Field Strengths vs. Source Type and Distance

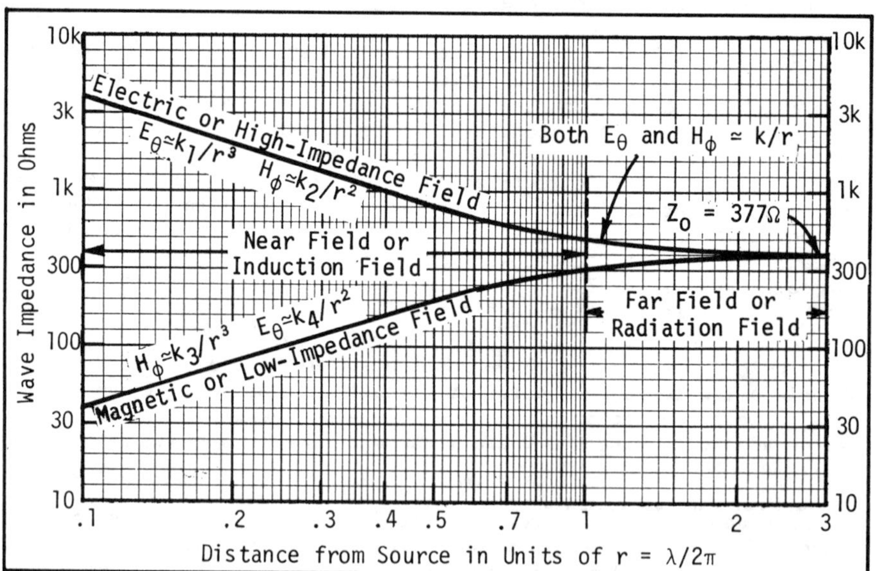

Figure 1.3 - Wave Impedance as a Function of Source Distance

in the far field. Fig. 1.3 also illustrates this wave impedance as a function of distance, r.

When the dimension, D, of an emission source (e.g., a wire or an antenna) becomes an appreciable fraction of a wavelength, the expressions presented in Eqs. (1.1) to (1.3) no longer apply. Different relations then must be used since the curvature of the arriving wave front no longer permits the incremental elements of an antenna or radiating source to be in phase. This situation is described in Chap. 2 of Vol. 2, "EMI Test Methods and Procedures".

1.2 WAVE IMPEDANCE

The previous section showed that the electromagnetic wave impedance is a function of the impedance of the source circuit generating the wave; the distance, r, from source-to-metal barrier measuring point; and the frequency. Summing up, then, the wave impedance, Z_w, is:

$$Z_w = E/H = kZ_o = k\sqrt{\mu/\varepsilon} = 120\pi k = 377k \text{ ohms} \tag{1.4}$$

where, $k \simeq \frac{\lambda}{2\pi r}$ for constant-current source (very high-impedance) and $r \leq \lambda/2\pi$ (1.5)

$\simeq \frac{2\pi r}{\lambda}$ for constant-voltage source (very low-impedance) and $r \leq \lambda/2\pi$ (1.6)

$\simeq 1$ for any network impedance and $r \geq \lambda/2\pi$ (1.7)

$\mu = \mu_o\mu_r$ = permeability of medium

μ_o = absolute permeability of air = $4\pi x10^{-7}$ henrys/m

μ_r = permeability of medium relative to air

$\varepsilon = \varepsilon_o\varepsilon_r$ = permitivity of medium

ε_o = absolute permitivity of air = $1/36\pi x10^9$ farads/m

ε_r = permitivity of medium relative to air

provided that the length, ℓ, of the source emission-generating circuit element is $\ell \ll \lambda$. Thus, in the far field the plane-wave impedance, 377 ohms, is independent of the circuit source impedance, but in the near field, the wave impedance can never exceed $377\lambda/2\pi r$ nor be less than 377 x $2\pi r/\lambda$.

The above raises the question about just how the source-generating circuit and wave impedances are related in the near field. This relation is important since it is shown in Sec. 1.5 that the shielding effectiveness in the near field is presented in terms of a high-impedance source (electric fields) and low-impedance source (magnetic fields). Use of electric-field sources may give rise to optimistic results (relatively high shielding effectiveness) and use of magnetic-field sources may give rise to pessimistic results (relatively low shielding effectiveness). The former could result in an underdesign condition and the latter could result in overdesign of any shielded circuit, box, equipment, cabinet, etc.

The development of a discrete relation between circuit, Z_c, and wave, Z_w, impedance in the near field is beyond the scope of this handbook. However, the following mathematical relations are suggested and all for conditions in which the circuit dimensions, $\ell \ll \lambda$:

$$Z_w \simeq \frac{Z_o\lambda}{2\pi r}, \text{ for } Z_c > \frac{Z_o\lambda}{2\pi r} \geq Z_o \quad (1.8)$$

$$\simeq Z_c, \text{ for } \frac{Z_o\lambda}{2\pi r} \geq Z_c \geq Z_o \quad (1.9)$$

$$\simeq Z_o, \text{ for } Z_c = Z_o \text{ or } \frac{\lambda}{2\pi r} \geq 1 \quad (1.10)$$

$$\simeq Z_c, \text{ for } Z_o > Z_c \geq \frac{Z_o 2\pi r}{\lambda} \quad (1.11)$$

$$\simeq \frac{Z_o 2\pi r}{\lambda}, \text{ for } Z_o > \frac{Z_o 2\pi r}{\lambda} > Z_c \quad (1.12)$$

Eqs. (1.8) to (1.12) are plotted in Fig. 1.4 for several values of Z_c including common transmission-line impedances of 50, 100, 300, and 600 ohms. To the extent that these conditions exist, the common transmission line impedances, then, never permit either a very high or very low wave impedance condition to exist when $r \ll \lambda/2\pi$.

As discussed later, sometimes it is more useful to present Eqs. (1.4) through (1.7) in terms of frequency, f, rather than wavelength. Since:

$$f\lambda_m = C = 1/\sqrt{\mu_o\varepsilon_o} = 3 \times 10^8 \text{ m/sec} \quad (1.13)$$

or,

$$f_{MHz} = 300/\lambda_m = \text{frequency in MHz} \quad (1.14)$$

where, f = frequency in Hz

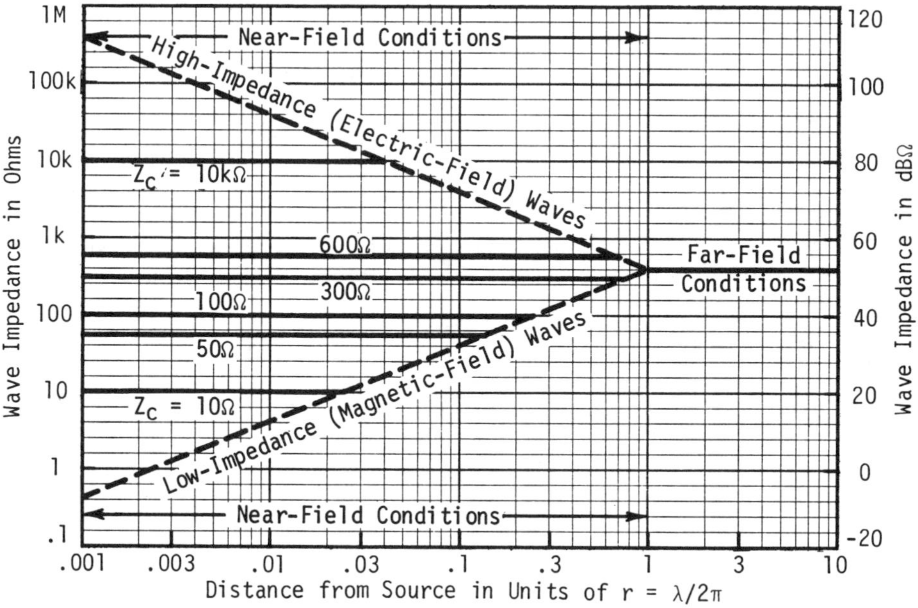

Figure 1.4 - Wave Impedance for Several Circuit Impedances

f_{MHz} = frequency in MHz

λ_m = wavelength in meters

C = velocity of light in air

Eqs. (1.4) through (1.7) then become:

High-Impedance Circuit in near field:

$$Z_w = \frac{Z_o\lambda}{2\pi r} = \frac{18{,}000}{r_m f_{MHz}} \text{ ohms} \tag{1.15}$$

Low-Impedance Circuit in near field:

$$Z_w = \frac{Z_o 2\pi r}{\lambda} = 7.9 r_m f_{MHz} \text{ ohms} \tag{1.16}$$

Far-field conditions:

$$Z_w = 120\pi = 377 \text{ ohms} \tag{1.17}$$

In terms of the more general conditions of any circuit impedance presented in Eqs. (1.8) through (1.12), the generated wave impedance is:

$$Z_w \simeq 18{,}000/r_m f_{MHz}, \text{ for } Z_c \geq 18{,}000/r_m f_{MHz} \tag{1.18}$$

$$\simeq Z_c, \text{ for } 18{,}000/r_m f_{MHz} \geq Z_c \geq 7.9 r_m f_{MHz} \tag{1.19}$$

$$\simeq 7.9 r_m f_{MHz}, \text{ for } 7.9 r_m f_{MHz} \geq Z_c \tag{1.20}$$

The foregoing relations will be used in interpreting the shielding effectiveness presented in the appendices.

1.3 METAL IMPEDANCE

All homogeneous materials are characterized by a quantity known as the intrinsic impedance of the material:

$$Z_i = \sqrt{\frac{j\omega\mu}{\sigma + j\omega\varepsilon}} \tag{1.21}$$

where, ω = $2\pi f$ radians

f = frequency in Hz

μ = permeability of the material = $\mu_o \mu_r$

μ_o = absolute permeability of air = $4\pi \times 10^{-7}$ henrys/meter

μ_r = permeability of material relative to air

σ = conductivity of material in mhos/meter

ε = permitivity of the material = $\varepsilon_o \varepsilon_r$

ε_o = absolute permitivity of air = $1/(36\pi \times 10^9)$ farads/meter

ε_r = permitivity of material relative to air

As an electromagnetic wave propagates through the material, the impedance of the wave, $Z_w = E/H$, approaches the value Z_i (see Eq (1.21)).

For air, conductivity is extremely small, i.e., $\sigma \ll \omega\varepsilon$. Thus, the intrinsic impedance of Eq. (1.21) for air becomes:

$$Z_a = \sqrt{\frac{\mu}{\varepsilon}} = \sqrt{\frac{\mu_o}{\varepsilon_o}} = \sqrt{4\pi \times 36\pi \times 10^2} = 120\pi = 377 \text{ ohms} \tag{1.22}$$

1.3.1 BARRIER IMPEDANCE OF METALS ($t >> \delta$)

In contrast to air and low-loss dielectrics, metals are defined as materials for which the conductivity is extremely high relative to air, such that $\sigma >> \omega\varepsilon$. Thus, the intrinsic impedance of Eq. (1.21) for a metal* becomes:

$$Z_m = \sqrt{\frac{j\omega\mu}{\sigma}} = \sqrt{\frac{j2\pi f\mu}{\sigma}} = (1+j)\sqrt{\frac{\pi f\mu}{\sigma}}, \text{ for } \sigma >> \omega\varepsilon \text{ and } t >> \delta^* \qquad (1.23)$$

$$|Z_m| = \sqrt{2} \times 10^3 \sqrt{\pi\mu f_{MHz}/\sigma} \; \Omega/\text{sq., for frequency in MHz} \qquad (1.24)$$

The intrinsic impedance of air (Eq. (1.22)) is a purely resistive constant, whereas the intrinsic impedance of a metal contains an equal resistive and inductive component. Consequently, Z_m depends upon both the permeability and conductivity of the metal.

Eq. (1.24) may be expressed in terms relative to copper:

$$|Z_m| = 369\sqrt{\mu_r f_{MHz}/\sigma_r} \quad \mu\Omega/\text{sq.} \qquad (1.25)$$

where, $\sigma = \sigma_c \times \sigma_r$

σ_c = conductivity of copper = 5.80×10^7 mhos/meter

σ_r = conductivity of metal relative to copper

Eq. (1.25) is plotted in Fig. 1.5 for various metals

The barrier impedance of metals is sometimes expressed in terms of skin depth, δ:

$$Z_m = \frac{(1+j)}{\sigma\delta} = \frac{\sqrt{2}}{\sigma\delta} \text{ (absolute value) ohms/sq.} \qquad (1.26)$$

The skin depth is defined as the surface thickness of a metal at any frequency for which 1-1/e or 63.2%** of the current is flowing therein. By combining Eqs. (1.23) and (1.26), the skin depth is defined:

$$\delta = \frac{\sqrt{2}}{\sigma Z_m} = \frac{1}{\sigma}\sqrt{\frac{\sigma}{\pi f\mu}} = \frac{1}{\sqrt{\pi f\mu\sigma}} \quad >> t \qquad (1.27)$$

---

* As described later, the surface impedance is based on a metal thickness which is very much greater than the skin depth, $t >> \delta$. As the frequency approaches zero, $\delta \to \infty$, and $Z_m \to 0$.

** Two skin depths = 86.5% and three skin depths = 95.0% of the total current flow. For 99% of the current flow, 4.6 skin depths are required. Often a metal thickness, t, is considered to be adequate when, $t > 3\delta$.

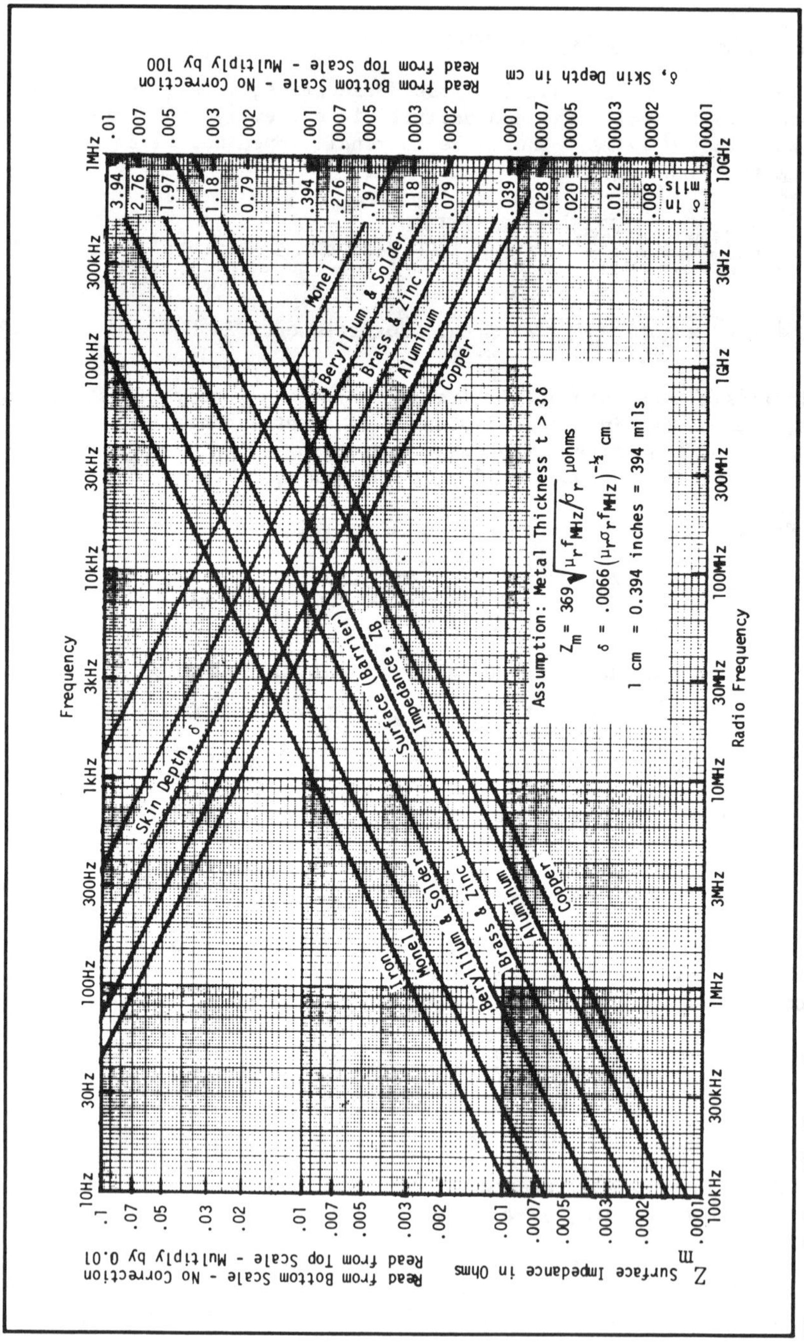

Figure 1.5 - Surface Impedance and Skin Depth of Various Metals vs. Frequency

$$= \sqrt{\frac{10^{-3}}{\pi f_{MHz}\mu\sigma}} \text{ for frequency in MHz} \tag{1.28}$$

$$= 0.0066/\sqrt{f_{MHz}} \quad \text{cm for copper } (\mu_r=\sigma_r=1) \tag{1.29}$$

$$= 2.6/\sqrt{f_{MHz}} \quad \text{mils* for copper } (\mu_r=\sigma_r=1) \tag{1.30}$$

$$= 0.0066/\sqrt{\mu_r\sigma_r f_{MHz}} \quad \text{cm for any metal} \tag{1.31}$$

$$= 2.6/\sqrt{\mu_r\sigma_r f_{MHz}} \quad \text{mils* for any metal} \tag{1.32}$$

Eqs. (1.31) and (1.32) are plotted in Fig. 1.5 for various materials.

1.3.2 BARRIER IMPEDANCE OF METALS ($t < 3\delta$)

In the previous section, the development and applications of Eqs. (1.24) and (1.25) are predicated upon the thickness, t, of the metal barrier being many skin depths, i.e., $t>>\delta$. In order to establish the applicability of the equations, a somewhat arbitrary situation is selected in which $t>3\delta$. When $t=3\delta$, 95% of the current flows between the surface and a depth of 3δ, and 5% of the current flows beyond a depth of 3δ. Thus, when $t=3\delta$, the barrier impedance must actually be 5% greater than indicated in Eqs. (1.24) and (1.25).

To accommodate any metal thickness, a generalization of the previous equations is needed, viz:

$$Z_B = \frac{Z_m}{\left(1-e^{-t/\delta}\right)}, \text{ for } t/\delta \text{ any value} \tag{1.33}$$

$$= Z_m \text{ for } t/\delta >> 1 \tag{1.34}$$

$$\simeq \frac{Z_m}{[1-(1-t/\delta)]}, \text{ for } t/\delta << 1 \tag{1.35}$$

$$\simeq \delta Z_m/t \text{ for } t/\delta << 1 \tag{1.36}$$

The error in Z_B in Eq. (1.36), which may develop from applying Z_m under conditions for which t/δ is not << 1 is:

$$\Delta Z_{dB} = -20 \log_{10}\left(1-e^{-t/\delta}\right) \text{ dB} \tag{1.37}$$

Eq. (1.37) is plotted in Fig. 1.6. The error is 4 dB when $t=\delta$ and approaches a 20 dB per decade slope with t/δ when t/δ >> 1.

---

* 1 mil = 0.001 inch = 0.00254 cm = 25.4 micron (or μm) = 25,400nm.

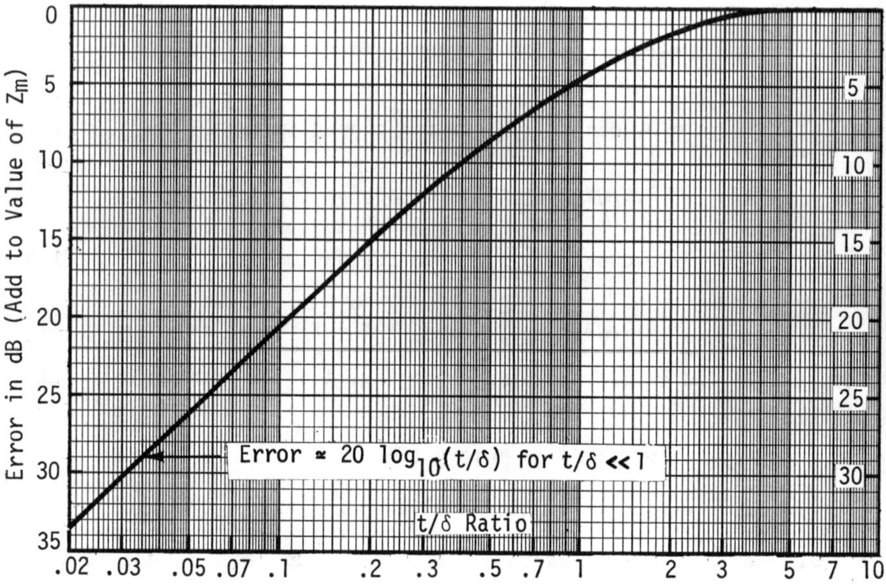

Figure 1.6 - Barrier Metal Impedance Error in Z_m Expression by Assuming $t/\delta \gg 1$

Substituting Eq. (1.33) in Eq. (1.25) yields the more general situation for Z_B of any thickness:

$$Z_B = 369\sqrt{\mu_r f_{MHz}/\sigma_r}\left(1-e^{-t/\delta}\right)^{-1} u\Omega/\text{sq.} \tag{1.38}$$

Eq. (1.38) is plotted in Fig. 1.7 for copper and iron corresponding to several values of t/δ. The $t/\delta \geq 3$ parameter line for copper in Fig. 1.7 is the same as the copper line in earlier Fig. 1.5.

Sometimes it is useful to know the barrier impedance corresponding to very thin deposits of metal on glass or clear plastic substrates such as used for electrically-conductive transparent EMI-shielded, optical-display windows. The conductively-coated windows may have deposits of much less than a micron thickness in which $t/\delta \ll 1$. For this situation Eq. (1.36) is directly obtained from Eq. (1.26):

$$Z_B = \delta Z_m/t = \frac{\sqrt{2}}{\sigma t}\ \Omega/\text{sq. for } t/\delta \ll 1 \tag{1.39}$$

$$= 244\text{x}10^{-8}/t_{cm}\ \Omega/\text{sq. for copper} \tag{1.40}$$

$$= 0.0244/t_{microns}\ \Omega/\text{sq. for copper} \tag{1.41}$$

$$= 96\text{x}10^{-5}/t_{mils}\ \Omega/\text{sq. for copper} \tag{1.42}$$

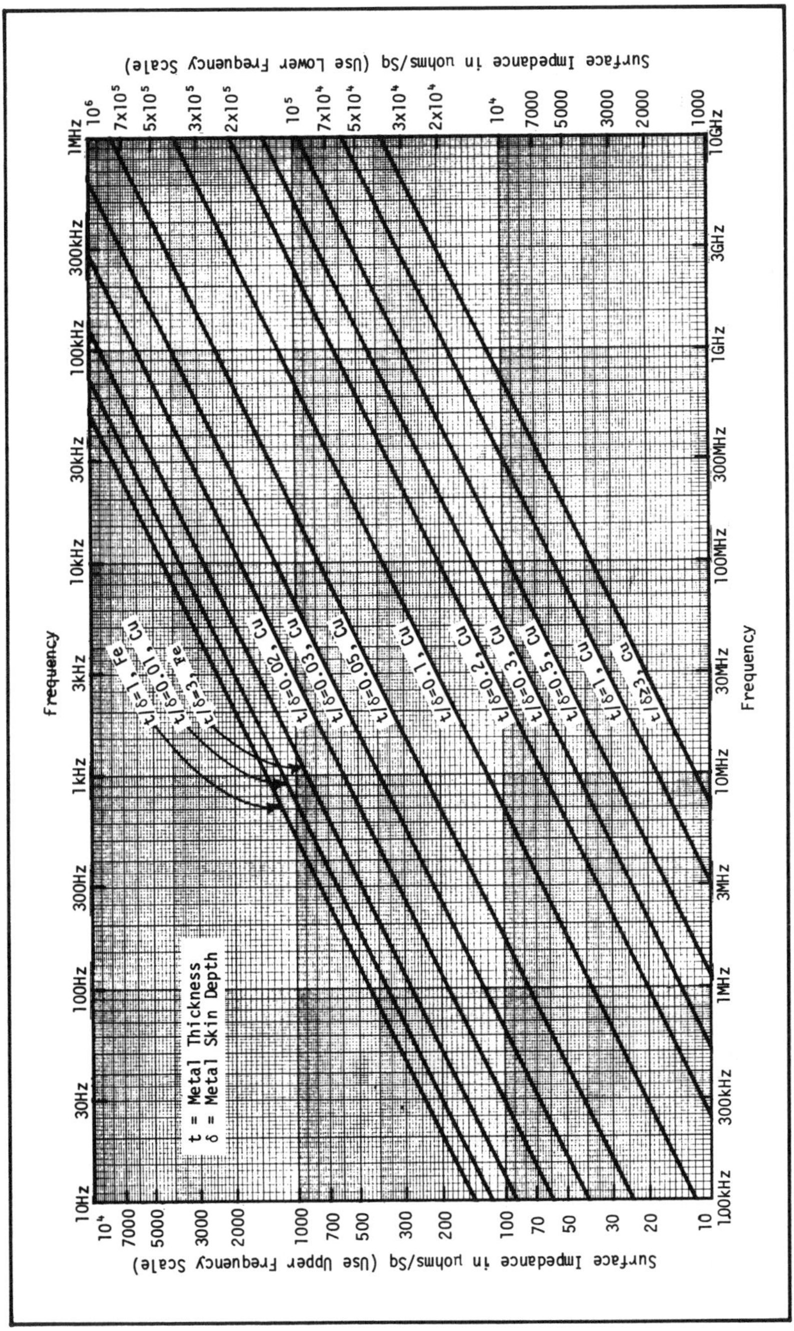

Figure 1.7 - Surface Impedance of Copper and Iron vs. Frequency and Skin Depth in Units of t/δ Ratios

It is noted that Eq. (1.39) is independent of the material permeability and frequency as long as $t/\delta << 1$. As discussed later, typical conductively-coated windows may have deposited metal corresponding to impedances of the order of 20 Ω/sq. and a light transmission of about 65% in the visible spectrum. Were copper* to be used as the deposited material, Eq. (1.40) indicates that a deposition of about .0012 microns or 1.22 nanometer would be required.

1.4 SHIELDING EFFECTIVENESS

This section describes a figure-of-merit for describing the performance of a shield in reducing the electromagnetic energy either impinging upon a potential susceptible victim or exiting from an undesirable source**. This figure-of-merit is called *shielding effectiveness* (SE) in the parlance of the Electromagnetic Compatibility (EMC) Community. It is defined as:

$$SE_{dB} = 10 \log_{10}\left(\frac{\text{incident power density}}{\text{transmitted power density}}\right) \tag{1.43}$$

where, incident power density = power density at a measuring point before shield is in place

transmitted power density = power density at same measuring point after shield is in place.

Eq. (1.43) is defined as a loss so that the shielding effectiveness is always positive.

In terms of the field strength, Eq. (1.43) may also be defined as follows as long as the two fields are measured in the same medium having the same wave impedance:

$$SE_{dB} = 20 \log_{10}\left(\frac{E_b}{E_a}\right) \text{ for electric fields} \tag{1.44}$$

$$= 20 \log_{10}\left(\frac{H_b}{H_a}\right) \text{ for magnetic fields} \tag{1.45}$$

where, E_b = electric-field strength *before* shield is installed

E_a = electric-field strength *after* shield is installed

---

* Because of the tendency to oxidize and lack of stability, copper is usually not used for conductively-coated windows. Sec. 2.1 discusses shielding effectiveness of conductively-coated windows using more stable metals.

** Reciprocity applies when the shield material is operated below saturation.

H_b = magnetic-field strength *before* shield is installed

H_a = magnetic-field strength *after* shield is installed

Eqs. (1.44) and (1.45) give identical results in the far field ($r \gg \lambda/2\pi$), but do not result in the same SE_{dB} in the near field since the wave impedances are different.

Fig. 1.8 shows the conceptual mechanism for determining shielding effectiveness. It is detailed in Fig. 1.9 and may be explained as follows. The incident field strength (an electric field is illustrated here) is considered as unity relative to itself. The reflected field is:

$$\rho = \frac{1-K}{1+K} = -1 \text{ for } K \gg 1 \tag{1.46}$$

$$= 0 \text{ for } K = 1 \tag{1.47}$$

$$= +1 \text{ for } 0 \geq K \ll 1 \tag{1.48}$$

where, $K = Z_w/Z_m$ (1.49)

Z_w = wave impedance defined in Sec. 1.2

Z_m = metal impedance defined in Sec. 1.3

and, $$\rho_{am} = \frac{1-K}{1+K} \tag{1.50}$$

= reflection coefficent at the air-to-metal interface (i.e., wave propagation from air medium into metal)

$$\rho_{ma} = \frac{K-1}{K+1} \tag{1.51}$$

= reflection coefficient at the metal-to-air interface (i.e., wave propagation from metal medium into air) in which case the roles of Z_w and Z_m are interchanged, whereby $K_{ma} = 1/K_{am}$ in Eq. (1.50) to Eq. (1.51)

The K term may be thought of as the voltage standing-wave ratio (VSWR) for $Z_w/Z_m \geq 1$ in Eq. (1.50). When $Z_w/Z_m \leq 1$, the VSWR concept still applies if Eq. (1.50) is defined as Eq. (1.51) in which $Z_m/Z_w \geq 1$.

The relative transmitted field, Γ_{am}, just inside the left edge of the metal-to-air barrier material is:

$$\Gamma_{am} = 1 - \rho_{am} \tag{1.52}$$

This field undergoes an attenuation in traversing the thickness of the metal barrier which turns the associated loss into exothermic

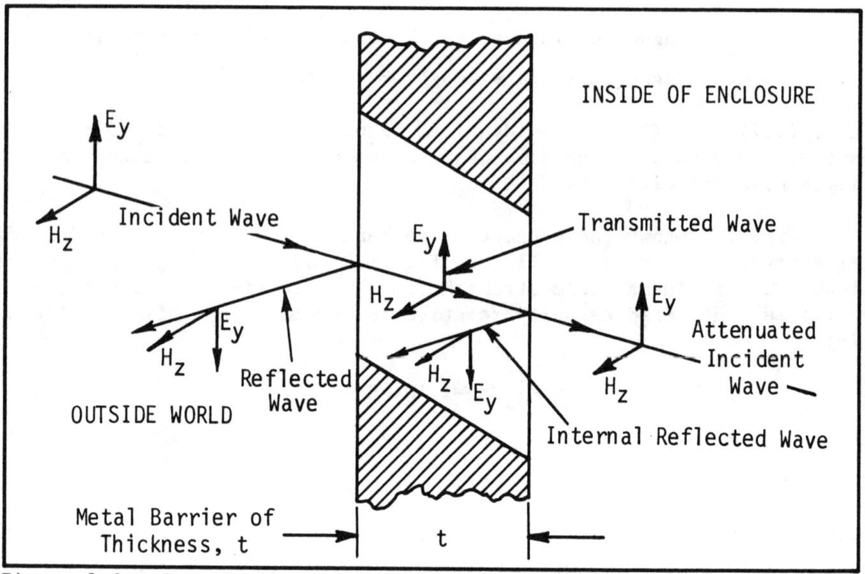

Figure 1.8 - Representation of Shielding Phenomena for Plane Waves

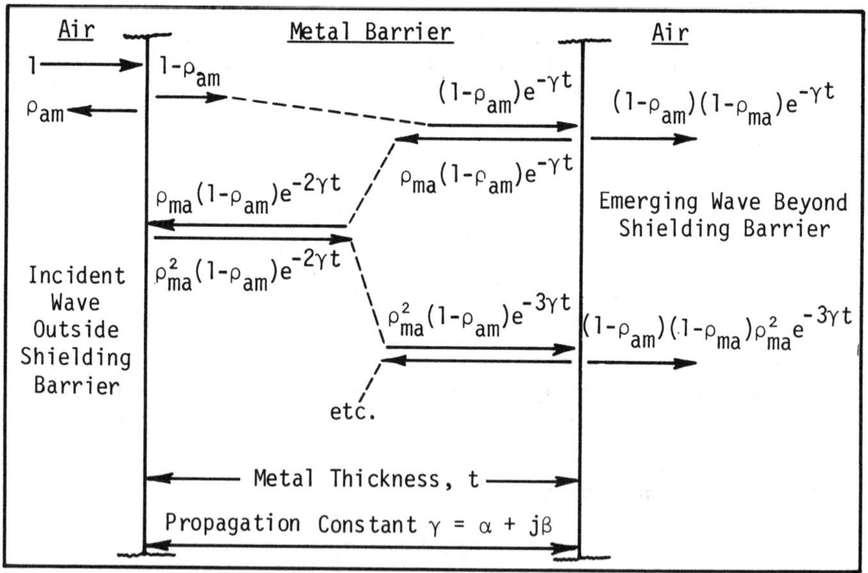

Figure 1.9 - Geometry of Metal Barrier Used in Explaining Shielding Effectiveness (See Text)

heat. The arriving field results in a lower field strength impinging at the right inside edge of the barrier:

$$\Gamma_{AR} = \Gamma_{am}e^{-\gamma t} = e^{-(\alpha+j\beta)t}\Gamma_{am} = e^{-(\alpha+j\beta)t}\left(1-\rho_{am}\right) \tag{1.53}$$

where, γ = propagation constant = $\alpha + j\beta$

α = attenuation constant

β = phase constant

t = metal thickness

The re-reflected relative field strength Γ_{RR}, at the inside right edge of the metal-to-air barrier of Fig. 1.9 is:

$$\Gamma_{RR} = \rho_{ma}\Gamma_{AR} = e^{-\gamma t}\rho_{ma}\left(1-\rho_{am}\right) \tag{1.54}$$

where, ρ_{ma} = metal-to-air reflection coefficient (Eq. (1.51)).

The relative transmitted field, Γ_{RT}, to the right just outside the metal barrier is:

$$\Gamma_{RT} = 1-\rho_{ma}\Gamma_{AR} = e^{-\gamma t}\left(1-\rho_{am}\right)\left(1-\rho_{ma}\right) \tag{1.55}$$

Eq. (1.55) is the shielding effectiveness expressed as a gain (a negative number) when $\gamma >> 1$.

When the propagation constant is not significant, one or more round-trip re-reflections must be considered. For example, the re-reflected field of Eq. (1.54) undergoes a second attenuation and phase shift in propagating back to the inside edge of the left metal-to-air interface of Fig. 1.9:

$$\Gamma_{LR} = \Gamma_{RR}e^{-\gamma t} = e^{-2\gamma t}\rho_{ma}\left(1-\rho_{am}\right) \tag{1.56}$$

The re-reflected field strength, Γ_{LL}, from the left edge inside barrier is:

$$\Gamma_{LL} = \rho_{ma}\Gamma_{LR} = e^{-2\gamma t}\rho_{ma}^2\left(1-\rho_{am}\right) \tag{1.57}$$

Undergoing a third attenuation and phase shift of this re-reflected field in arriving back at the inside face of the right edge, the relative field strength, Γ_{AR}, becomes:

$$\Gamma'_{AR} = \Gamma_{LL}e^{-\gamma t} = 3^{-3\gamma t}\rho_{ma}^2\left(1-\rho_{am}\right) \tag{1.58}$$

Finally, the transmitted component of Γ'_{AR} to the right outside the metal barrier is:

$$\Gamma'_{RT} = \left(1-\rho_{ma}\right)\Gamma'_{AR} = e^{-3\gamma t}\rho_{ma}^2\left(1-\rho_{am}\right)\left(1-\rho_{ma}\right) \tag{1.59}$$

Since the re-reflected field component of Eq. (1.59) is coherent with the direct transmitted field of Eq. (1.55), they are coherently added:

$$\Gamma_T = e^{-\gamma t}\left(1-\rho_{am}\right)\left(1-\rho_{ma}\right)\Big[1 + \underbrace{e^{-2\gamma t}\rho_{ma}^2}_{\text{First Round-trip Re-Reflections}} + \underbrace{e^{-4\gamma t}\rho_{ma}^4 + \ldots}_{\text{Multiple Round-trip Re-Reflections}}\Big] \tag{1.60}$$

The terms in the bracket constitute an infinite series (i.e., an infinity of re-reflections). The bracket expression can be simplified by writing this series in terms of its reciprocal. Thus, Eq. (1.60) becomes:

$$\Gamma_T = e^{-\gamma t}\left(1-\rho_{am}\right)\left(1-\rho_{ma}\right)\left(1-\rho_{ma}^2 e^{-2\gamma t}\right)^{-1} \tag{1.61}$$

Eq. (1.61) may be expressed in terms of the impedance ratio of the air-metal and metal-air interfaces by substituting Eqs. (1.50) and (1.51) therein:

$$\Gamma_T = e^{-\alpha t}\left(\frac{2K}{1+K}\right)\left(\frac{2}{1+K}\right)\left[1 - \left(\frac{K-1}{K+1}\right)^2 e^{-2\gamma t}\right]^{-1} \tag{1.62}$$

$$= \underbrace{e^{-\alpha t}}_{\text{Absorption Term (A)}}\ \underbrace{\frac{4K}{(1+K)^2}}_{\text{Reflection Term (R)}}\ \underbrace{\left[1 - \left(\frac{K-1}{K+1}\right)^2 e^{-2\gamma t}\right]^{-1}}_{\text{Re-Reflection Correction Term (B)}} \tag{1.63}$$

Expressing Eq. (1.63) as a loss (i.e., shielding effectiveness) rather than a gain, and converting it to decibels, there results:

$$SE_{dB} = 20\log_{10}\left(1/\Gamma_T\right) = 20\log_{10}\left\{e^{\alpha t}\left|\frac{(1+K)^2}{4K}\right.\left[1-\left(\frac{K-1}{K+1}\right)^2 e^{-2\gamma t}\right]\right|\right\} \tag{1.64}$$

where,

$$\text{Absorption Loss, } A_{dB} = 8.686\alpha t \tag{1.65}$$

$$\text{Reflection Loss, } R_{dB} = 20\log_{10}(1+K)^2/4K \tag{1.66}$$

$$\text{Re-Reflection Correction, } B_{dB} = 20\log_{10}\left|1-\left\{(K-1)^2/(K+1)^2\right\}e^{-2\gamma t}\right| \tag{1.67}$$

Eq. (1.64) is plotted in graphs in the appendices in this handbook for several metals, metal properties, metal thicknesses, distances, and frequencies. This question will now be examined in further detail of its three loss components: (1) absorption loss, (2) reflection loss, and (3) re-reflection loss correction.

1.4.1 ABSORPTION LOSS

Eq. (1.65) may be expanded:

$$A_{dB} = 8.686\alpha t = 8.686t\sqrt{\pi f\mu\sigma}$$

where, $\gamma = \alpha + j\beta = \sqrt{j\omega\mu(\sigma+j\omega\varepsilon)}$ (1.68)

$= \sqrt{j\omega\mu\sigma}$ since $\sigma >> \omega\varepsilon$ for metals (1.69)

$= (1+j)\sqrt{\pi f\mu\sigma}$ (1.70)

or, $\alpha = \beta = \sqrt{\pi f\mu\sigma}$ for metals (1.71)

If Eq. (1.68) is defined in terms of t in mils (thousandths of an inch) and f in MHz, and for t in cm it becomes for both the English and metric system of units:

$A_{dB} = 3.338t_{mils}\sqrt{f_{MHz}\mu_r\sigma_r}$ dB, English units (1.72)

$= 1314.3t_{cm}\sqrt{f_{MHz}\mu_r\sigma_r}$ dB, metric units (1.73)

where, μ_r and σ_r are permeability and conductivity relative to copper

Eqs. (1.72) and (1.73) are plotted in Figs. 1.10 through 1.17 for various metals: copper, aluminum, brass, beryllium, monel, iron, stainless steel, and the exotic high-permeability metals*.

1.4.2 REFLECTION LOSS

The reflection loss relations are predicated upon an impedance mismatch at the metal-barrier interfaces. Thus, it is useful to substitute the impedances of Eq. (1.49) by their equivalents from Eq. (1.4) for Z_w and Eq. (1.23) for Z_m:

$$K = \frac{Z_w}{Z_m} = \frac{k\sqrt{\mu_o/\varepsilon_o}}{(1+j)\sqrt{\pi f\mu/\sigma}} \quad (1.74)$$

where, $k = \lambda/2\pi r = 1/2\pi rf\sqrt{\mu_o\varepsilon_o}$ for high-impedance, E fields (1.75)

$k = 2\pi r/\lambda = 2\pi rf\sqrt{\mu_o\varepsilon_o}$ for low-impedance, H fields (1.76)

$= 1$ for far fields, $r \geq \lambda/2\pi$ (1.77)

Combining Eq. (1.74) through (1.77) yields:

* For magnetic materials (μ_r>1), the graphs are accurate only for the stipulated μ_r condition which varies with both magnetic-field strength (or flux density) and frequency, especially above several kHz (see Sec. 2.1.4).

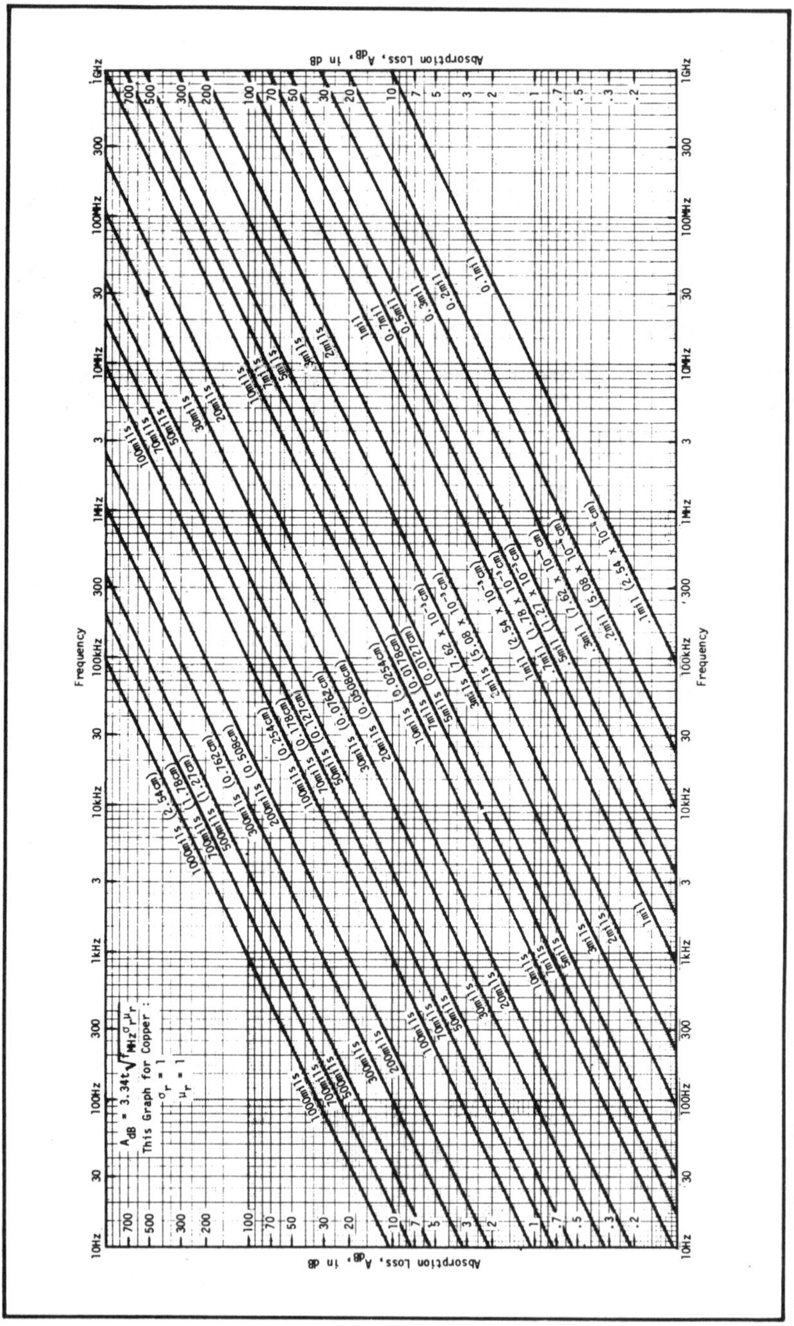

Figure 1.10 - Absorption Loss vs. Frequency and Thickness for Copper

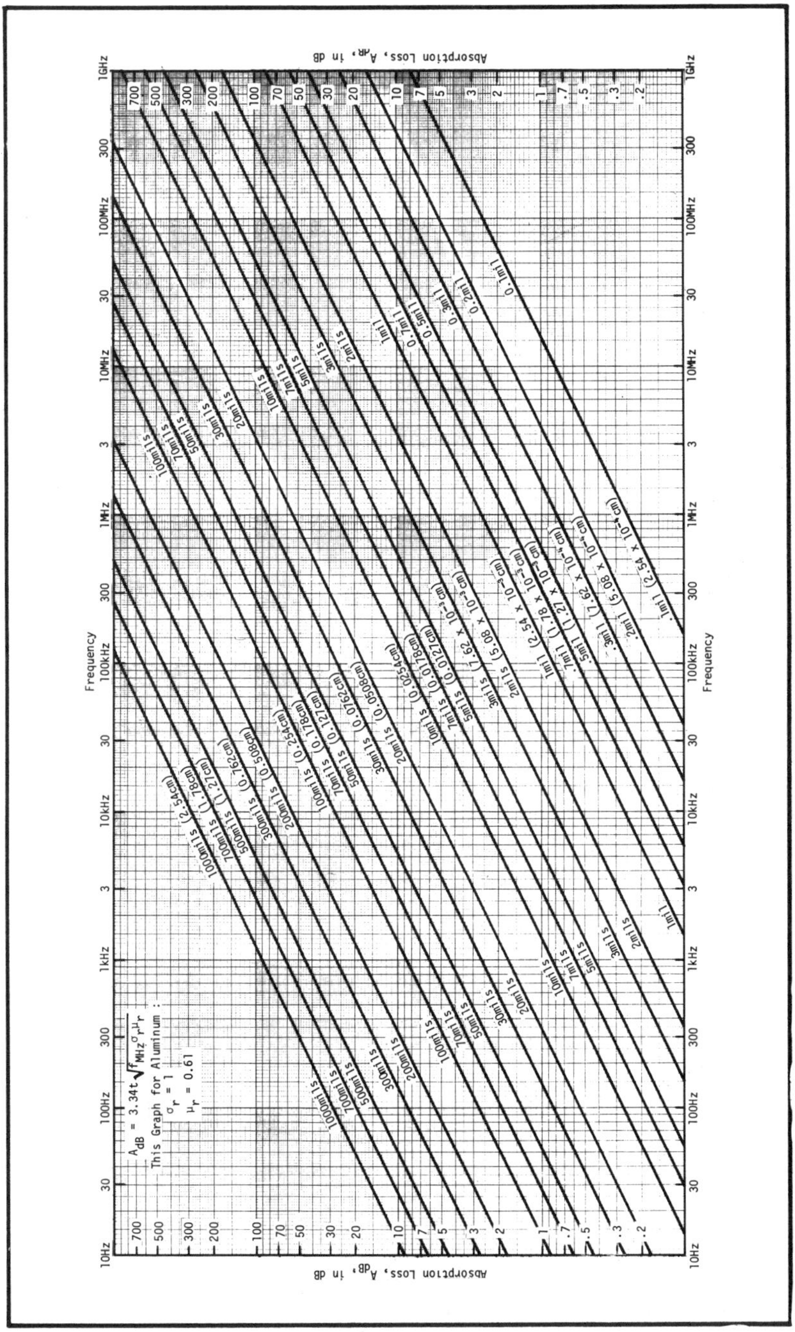

Figure 1.11 - Absorption Loss vs. Frequency and Thickness For Aluminum

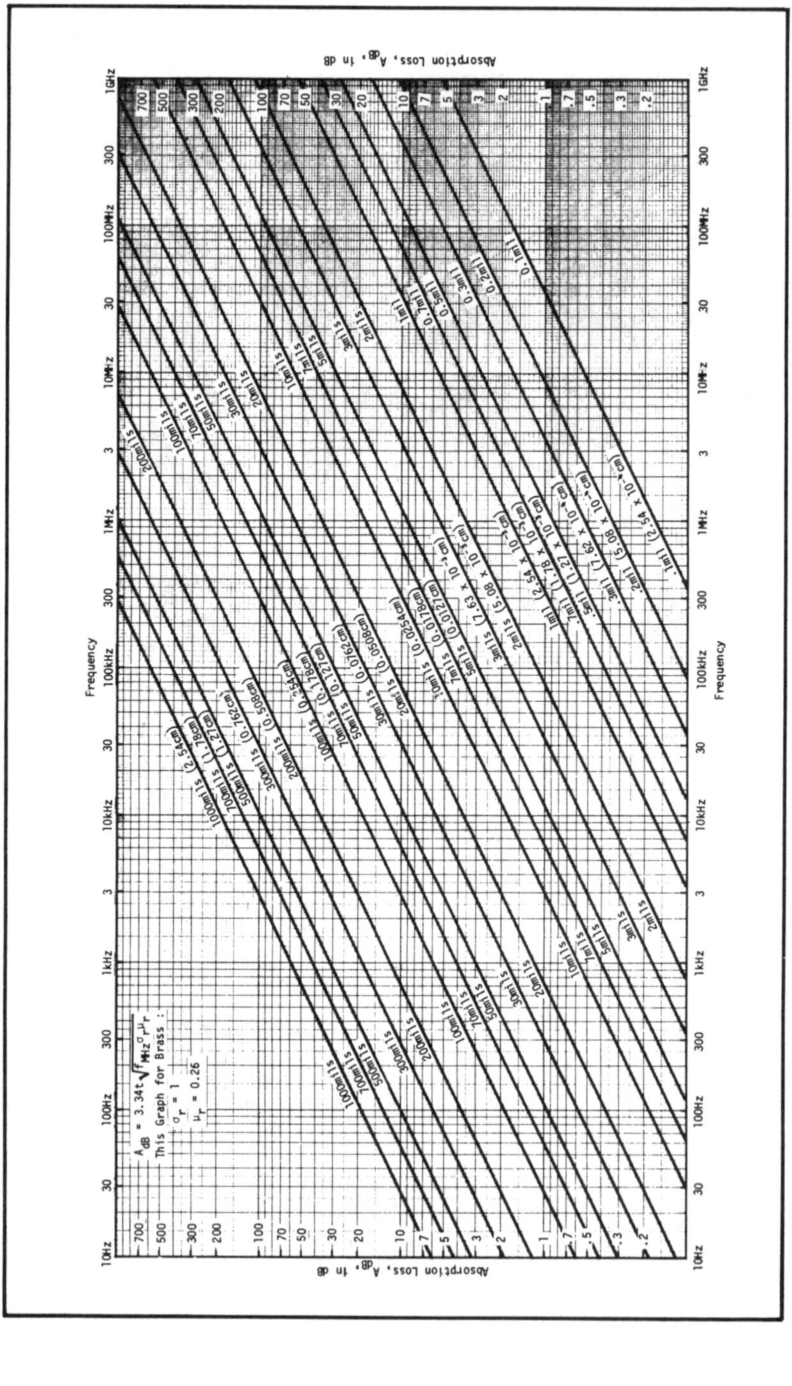

Figure 1.12 - Absorption Loss vs. Frequency and Thickness for Brass

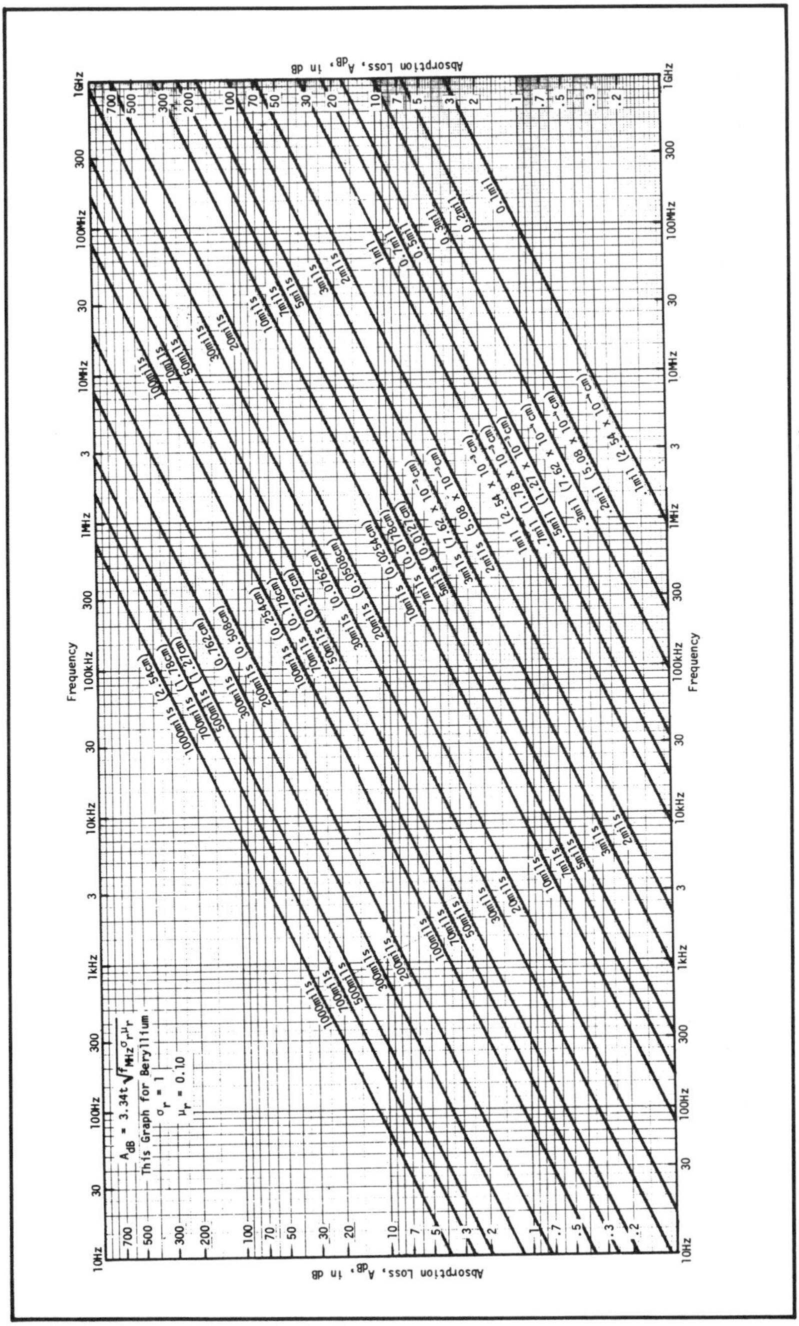

Figure 1.13 - Absorption Loss vs. Frequency and Thickness for Beryllium

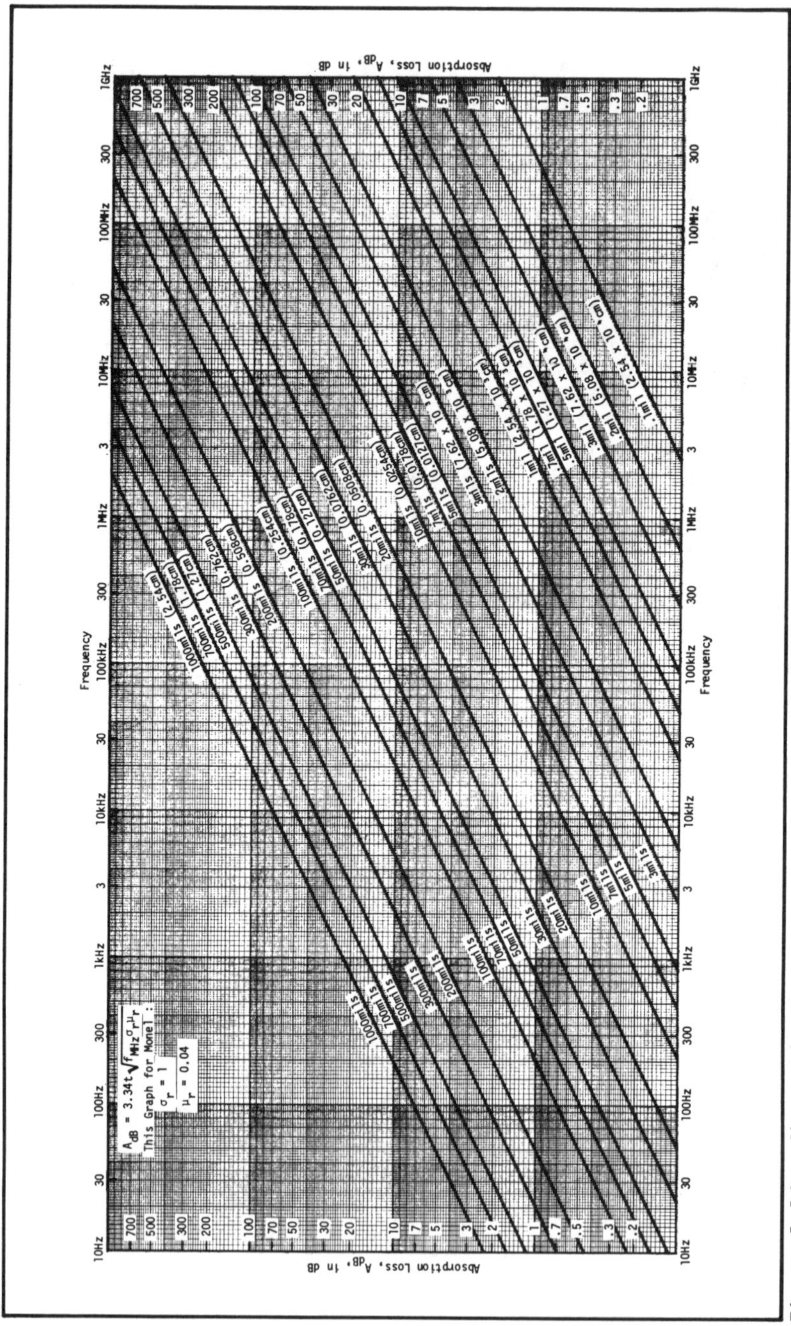

Figure 1.14 - Absorption Loss vs. Frequency and Thickness for Monel

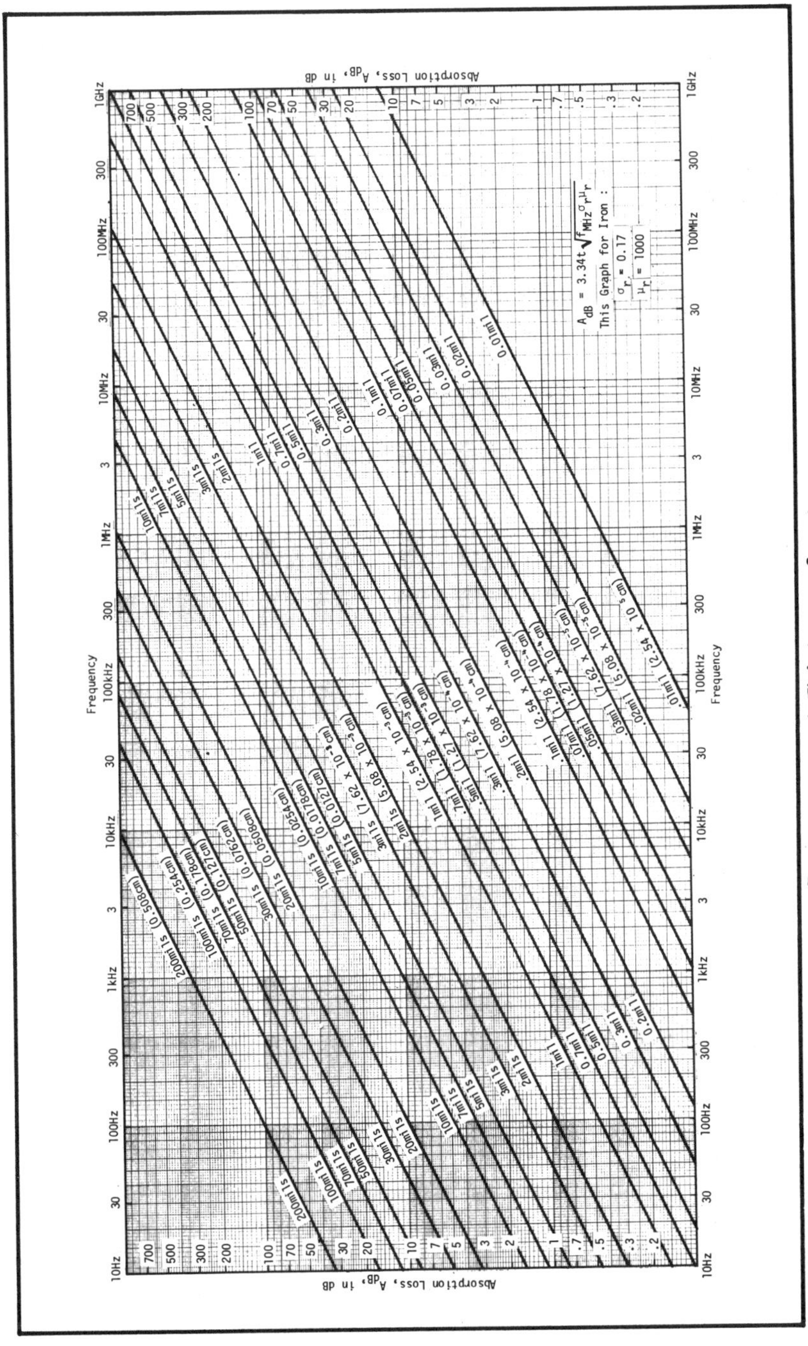

Figure 1.15 - Absorption Loss vs. Frequency and Thickness for Iron

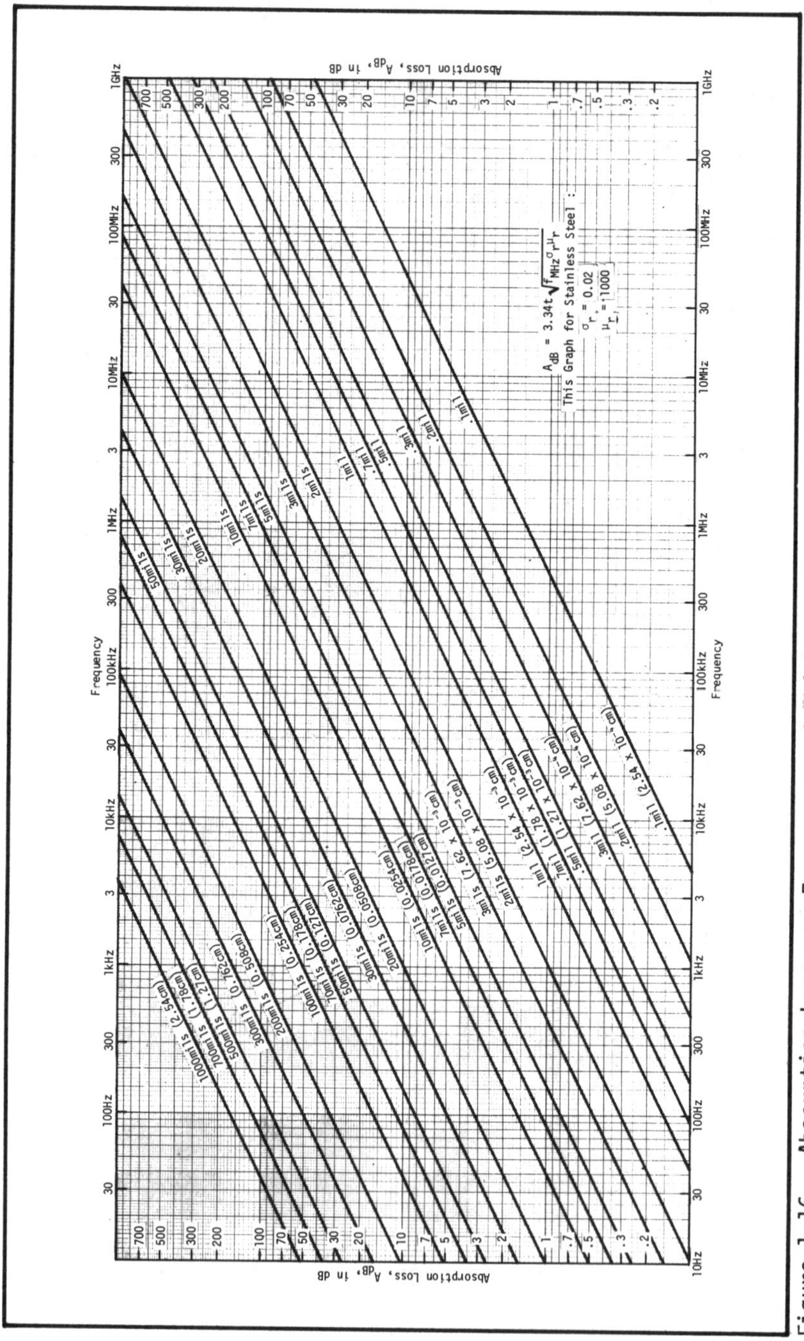

Figure 1.16 - Absorption Loss vs. Frequency and Thickness for Stainless Steel

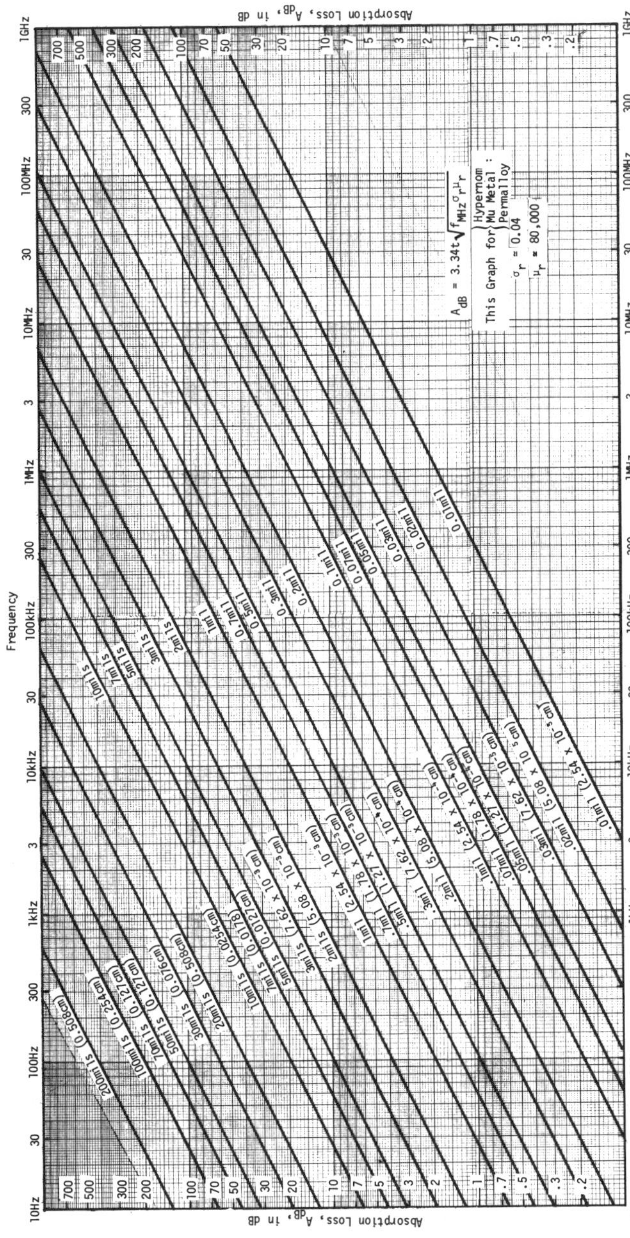

Figure 1.17 - Absorption Loss vs. Frequency and Thickness for High-Permeable Metals

$$K = 1/2\pi r f \varepsilon_o \sqrt{2\pi f \mu/\sigma} \quad \text{for high-impedance fields} \tag{1.78}$$

$$= r\sqrt{2\pi f \sigma \mu_o/\mu_r} \quad \text{for low-impedance fields} \tag{1.79}$$

$$= 1/\sqrt{2\pi f \mu_r \varepsilon_o/\sigma} \quad \text{for plane waves} \tag{1.80}$$

The reflection loss term of Eq. (1.66) may be expanded in terms of the wave and metal barrier impedances:

$$R_{dB} = 20 \log_{10}(1+K)^2/4K \tag{1.81}$$

$$\simeq 20 \log_{10}(K/4) = 20 \log_{10}\left(Z_w/4Z_m\right) \quad \text{for } K >> 1 \tag{1.82}$$

Eqs. (1.78) - (1.80) may be substituted into Eq. (1.82) to yield:

Electric (High-Impedance) Fields:

$$R_{dB} \simeq 152.0 - 10 \log_{10}\left(\mu_r f_{MHz}^3 r_{ft}^2/\sigma_r\right) \text{dB, English units} \tag{1.83}$$

$$\simeq 141.7 - 10 \log_{10}\left(\mu_r f_{MHz}^3 r_m^2/\sigma_r\right) \text{dB, metric units} \tag{1.84}$$

where, r_{ft} = source-to-barrier distance in feet

r_m = source-to-barrier distance in meters

Magnetic (Low-Impedance) Fields:

$$R_{dB} \simeq 64.3 - 10 \log_{10}\left(\mu_r/f_{MHz}\sigma_r r_{ft}^2\right) \text{ dB, English units} \tag{1.85}$$

$$\simeq 74.6 - 10 \log_{10}\left(\mu_r/f_{MHz}\sigma_r r_m^2\right) \text{ dB, metric units} \tag{1.86}$$

Plane Waves (Far-Field Conditions):

$$R_{dB} = 108.1 - 10 \log_{10}\left(\mu_r f_{MHz}/\sigma_r\right) \quad \text{dB, both units} \tag{1.87}$$

(independent of r as long as $r >> \lambda/2\pi = 47.8/f_{MHz}$)

When the concept of circuit impedance is introduced rather than simply high or low-impedance waves (cf. Eqs. (1.8) through (1.20), the shielding effectiveness in the near-field becomes:

$$R_{dB} \simeq 20 \log_{10}\left(\frac{Z_c}{4\text{x}369\text{x}10^{-6}\sqrt{\mu_r f_{MHz}/\sigma_r}}\right) \tag{1.88}$$

$$\left.\begin{aligned} &= 56.6 - 10 \log_{10}\left(\mu_r f_{MHz}/\sigma_r Z_c^2\right) \\ &\text{or} = \text{Eqs. (1.83) and (1.84)}\end{aligned}\right\} \text{whichever is less for E fields} \tag{1.89}$$

$$\left.\begin{aligned} &= 56.6 - 10 \log_{10}\left(\mu_r f_{MHz}/\sigma_r Z_c^2\right) \\ &\text{or} = \text{Eqs. (1.85) and (1.86)}\end{aligned}\right\} \text{whichever is less for H fields} \tag{1.90}$$

= Eq. (1.87) for plane waves.

For circuits whose impedance $Z_c \simeq 377$ ohms, the reflection loss appears to be that of far-field conditions, even if r is in the near field.

1.4.3 RE-REFLECTION CORRECTION

Eq. (1.67) may be expanded in terms of the wave and metal-barrier impedance (see Eq. (1.74)):

$$B_{dB} = 20 \log_{10} \left| 1 - \left((K-1)^2/(K+1)^2\right) e^{-2\gamma t} \right| \quad (1.91)$$

$$= 20 \log_{10} \left| 1 - \left(\frac{K-1}{K+1}\right)^2 10^{-.1A_{dB}} \left(\cos 0.23A_{dB} - j\sin 0.23A_{dB}\right) \right| \quad (1.92)$$

$$\simeq 20 \log_{10} \left(1 - e^{-2(\alpha+j\beta)t}\right), \text{for } K >> 1 \quad (1.93)$$

$$\simeq 20 \log_{10} \left(1 - e^{-2t\sqrt{\pi f \mu \sigma}} e^{-j2t\sqrt{\pi f \mu \sigma}}\right) \quad (1.94)$$

where, A_{dB} is given in Eqs. (1.72) and (1.73)

Eq. (1.92) is plotted in Fig. 1.18.

1.4.4 TOTAL LOSSES FOR K >> 1

It may be a bit misleading to think of a reflection loss and re-reflection correction as separate terms. After all, a reflection loss should be the entire loss including re-reflection.

For certain conditions, Eq. (1.64) may be greatly simplified:

$$SE_{dB} = 20 \log_{10} \left\{ e^{t/\delta} \left| \frac{(K+1)^2}{4K} \left[1 - \left(\frac{K-1}{K+1}\right)^2 e^{-2\gamma t}\right] \right| \right\} \quad (1.64)$$

where,

$$1/\delta = \gamma/\sqrt{2} = \alpha = \sqrt{\pi f \mu \sigma} \quad (1.95)$$

For conditions in which a substantial metal barrier impedance and wave impedance mismatch exists, i.e., for K >> 1, Eq. (1.64) becomes:

$$SE_{dB} \simeq 20 \log_{10} \left[e^{t/\delta} \left| \frac{K}{4} \left(1 - e^{-2\gamma t}\right) \right| \right] \text{ for } K >> 1 \quad (1.96)$$

$$= 20 \log_{10} \left[\underbrace{e^{t/\delta}}_{\text{Absorption Loss}} \underbrace{\left| \frac{K}{4} \left(1 - e^{-2t/\delta} e^{-j2t/\delta}\right) \right|}_{\text{Total Reflection Loss}} \right] \text{ for } K >> 1 \quad (1.97)$$

When γt (or t/δ) is very small (i.e., $\gamma t << 1$), the exponential term in Eq. (1.96) may be expanded in a power series and the first two

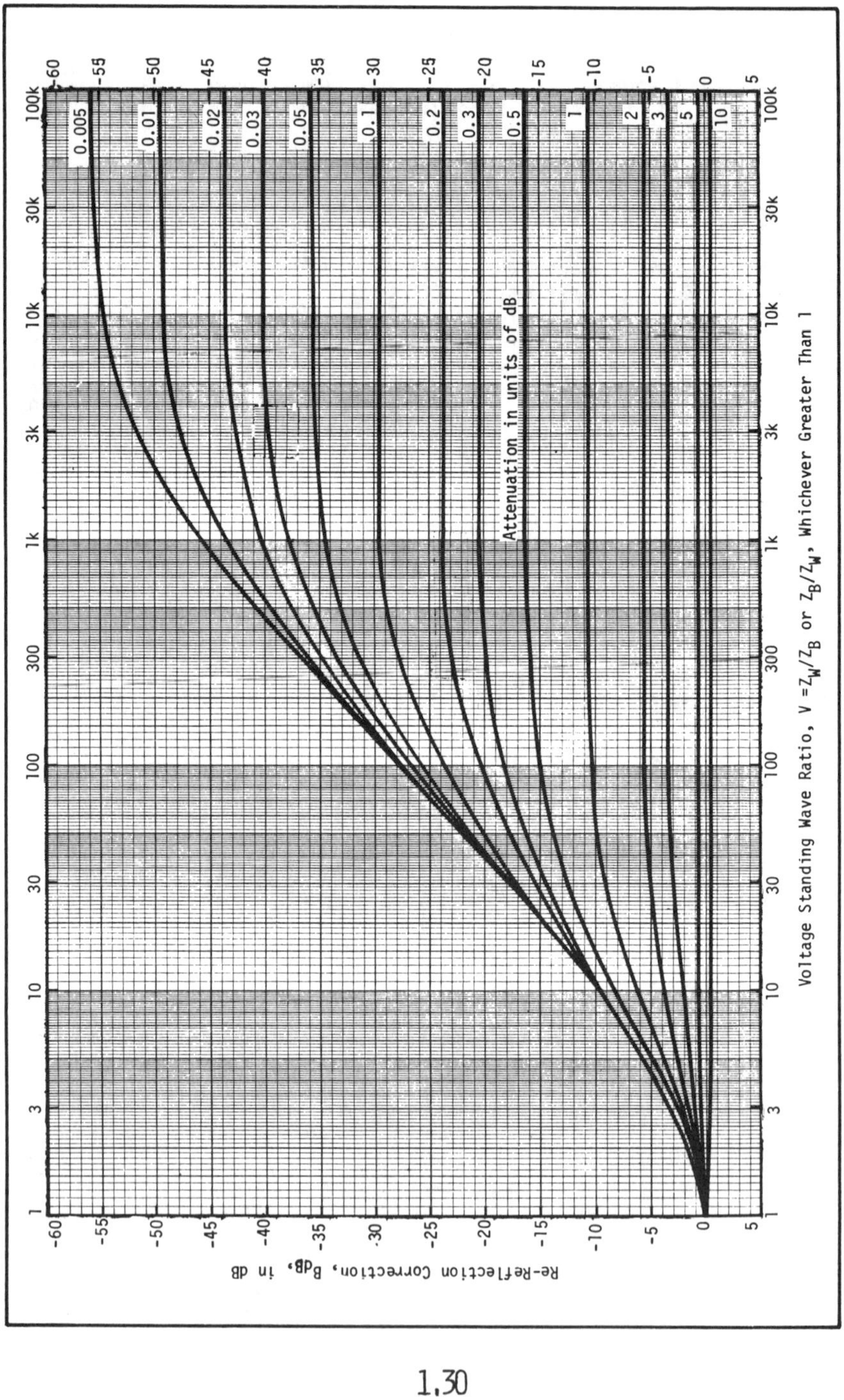

Figure 1.18 - Re-Reflection Correction vs. VSWR and Material Absorption Loss

terms retained. This results in:

$$SE_{dB} \simeq 20 \log_{10}\left[e^{t/\delta}\left|\frac{K}{4}(2\sqrt{2}t/\delta)\right|\right] \quad (1.98)$$

$$\simeq 20 \log_{10}\left[(1+t/\delta)\ \frac{K\sqrt{2}t/\delta}{2}\right] \quad (1.99)$$

$$\simeq 20 \log_{10}(0.707Kt/\delta),\ \text{for } t/\delta<<1 \text{ and } K>>1 \quad (1.100)$$

The t/δ term of Eq. (1.100) may be compared with Eq. (1.35) involving the metal impedances for t/δ << 1. Eq. (1.100) applies only when the metal thickness is a small fraction of its skin depth at any frequency.

On the other hand, when t/δ is >> 1, Eq. (1.97) becomes:

$$SE_{dB} = 20 \log_{10}\left(Ke^{t/\delta}/4\right),\ \text{for } t/\delta>>1 \text{ and } K>>1 \quad (1.101)$$

Eqs. (1.100) and (1.101) become equal when $2\sqrt{2}t/\delta = e^{t/\delta}$ or when $t/\delta = 0.74$. This may be considered to be a *corner-condition frequency*.

Substituting Eq. (1.27) into Eq. (1.100) yields:

$$SE_{dB} \simeq 20 \log_{10}(188.5\ k\sigma t),\ \text{for } t/\delta < 0.74 \text{ and } K \geq 10 \quad (1.102)$$

$$\simeq 20 \log_{10}\left(188.5\ kt/R_s\delta\right) = 20 \log_{10}(188.5\ k/R_{sdc})^* \quad (1.103)$$

$$= 108.9 + 20 \log_{10}\left(k\sigma_r t_{mils}\right)\ \text{dB, English system} \quad (1.104)$$

$$= 160.8 + 20 \log_{10}\left(k\sigma_r t_{cm}\right)\ \text{dB, metric system} \quad (1.105)$$

Substituting Eq. (1.27) into Eq. (1.101) yields:

$$SE_{dB} = 20 \log_{10}\left(66.6\ ke^{t/\delta}/R_s\right) \quad (1.106)$$

$$= 8.686t\sqrt{\pi f\mu\sigma} + 20 \log_{10}(66.6k/\sqrt{\pi f\mu/\sigma})\ \text{for } t/\delta>0.74 \text{ and } K\geq 10 \quad (1.107)$$

$$=108.1 + 1314.4t_{cm}\sqrt{f_{MHz}\mu_r\sigma_r} + 20 \log_{10}\left(k/\sqrt{f_{MHz}\mu_r/\sigma_r}\right) \quad (1.108)$$

where, $k = 1$ for far fields ($r \geq \lambda/2\pi$)
$= \lambda/2\pi r$ for E fields
$= 2\pi r/\lambda$ for H fields
R_s = surface resistivity in ohms/sq
$= Z_m/(1+j) = 1/\sigma\delta$

To facilitate which equation to use, $t/\delta = 0.74$ may be obtained

* $R_{sdc} = 1/\sigma t$ at DC = D-C surface resistivity (1.108A)

from δ of Fig. 1.5 at any frequency for the metals used. Eq. (1.106) is plotted in Fig. 1.19 for values of R_s ranging from 10mΩ to 30Ω, k from emission-source-to-metal-barrier distance values of $10^{-5}\lambda/2\pi$ to $100\lambda/2\pi$ and $t/\delta = 0.74$. The shielding effectiveness will ordinarily be greater than that shown in Fig. 1.19 because of absorption loss provided $t/\delta > 0.74$. Note that for all values of k, S_{dB} is basically a reflection loss, viz, an impedance mismatch between the wave impedance and the metal impedance. Thus, the object is to economically achieve a low metal resistance using foils, conductive paints, flame-spray metals, meshes, screens, etc. as explained in Chap. 2.

1.4.5 LOW-FREQUENCY MAGNETIC SHIELDING EFFECTIVENESS

Eq. (1.64) presented the overall shielding effectiveness expression applicable for far field and electric and magnetic fields in the near field for any thickness of any metal. A substantial simplification of Eq. (1.64) results at and below very-low frequency for magnetic materials when t/δ, K, and $2\gamma t \ll 1$:

$$SE_{dB} \simeq 20 \log_{10}\left|\frac{1}{4K}\left[1 - (1-4K)(1-2\gamma t)\right]\right| \tag{1.109}$$

$$\simeq 20 \log_{10}\left|\frac{1}{4K}\left[1 - (1-4K-2\gamma t)\right]\right| \tag{1.110}$$

for $8K\gamma t \ll 4K$ and $2\gamma t$

$$\simeq 20 \log_{10}(1 + \gamma t/2K) \tag{1.111}$$

where, $\gamma = (1+j)\alpha = \sqrt{2\pi f\mu\sigma}$ (1.68)

$K = Z_w/Z_m = r\sqrt{2\pi f\sigma\mu_o/\mu_r}$ for magnetic fields (1.79)

Substituting Eqs. (1.68) and (1.79) into Eq. (1.111) yields:

$$SE_{dB} \simeq 20 \log_{10}\left(1 + \frac{t\sqrt{2\pi f\mu_o\mu_r\sigma}}{2r\sqrt{2\pi f\sigma\mu_o/\mu_r}}\right) \tag{1.112}$$

$$\simeq 20 \log_{10}\left(1 + \frac{\mu_r t}{2r}\right), \text{ for } t/\delta, \text{ K and } 2\gamma t \ll 1 \tag{1.113}$$

Eq. (1.113) is a useful and simple expression for very low-frequency and D-C shielding effectiveness of magnetic fields. It is equally useful to determine the shielding effectiveness of a wire pair, coax, or harness having a magnetic shield of thickness, t, and radius, r. If a wire or harness shield is used, be sure that the *effective* relative permeability, μ_r, is used for braided shields since much of the shield may be air or other material. Eq. (1.113) is plotted in Fig. 1.20.

<u>Example</u>: Determine the low-frequency magnetic shielding effectiveness

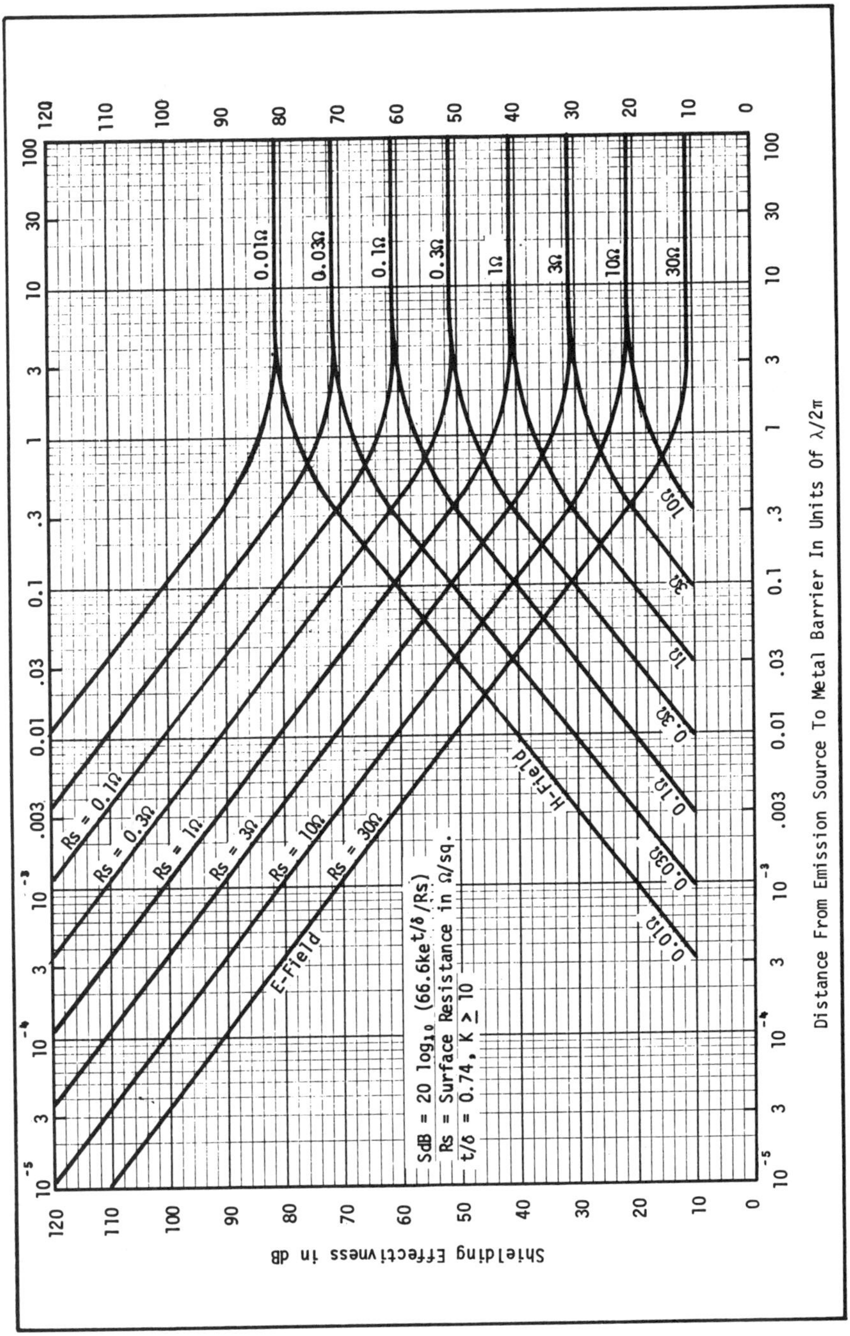

Figure 1.19 - Shielding Effectiveness vs. Metal-to-Emission Distance and Surface Resistances

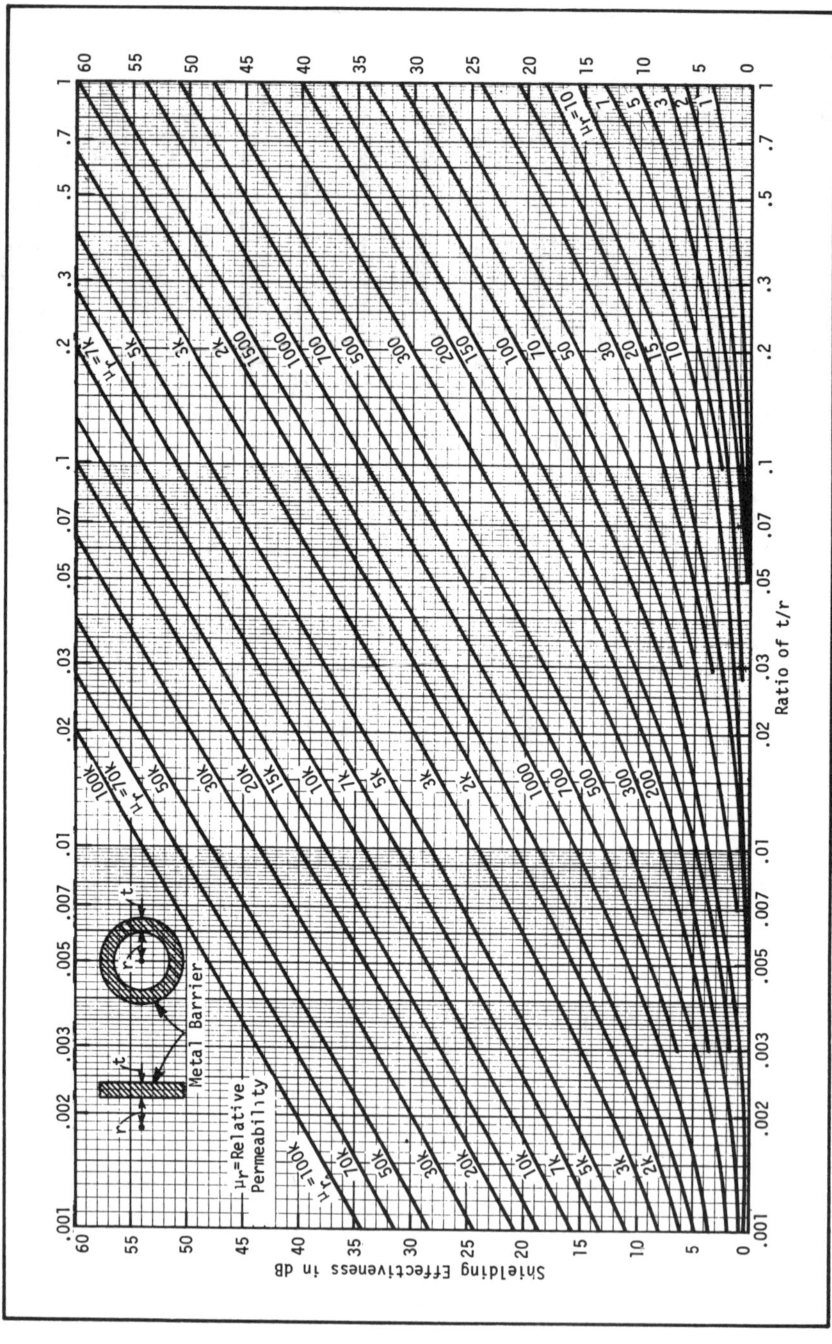

Figure 1.20 - Low Frequency, Shielding Effectiveness to Magnetic Fields

of a tin-plated, copper-clad steel braid in which tin = 3%, copper = 37%, and steel = 60%. The braid is 80% material and 20% air by volume and has an average thickness of t = 0.020 inches. The braided shield, when slipped over a harness has an average radius of 0.50 inches:

$$\mu_r = M_{de}\, MM_{de}\mu_{MM} + NMM_{de}\text{x}1 + A_{de}\text{x}1 \qquad (1.114)$$

$$\simeq M_{de}MM_{de}\mu_{MM}, \text{ since } NMM_{de} << MM_{de}\mu_{MM} \qquad (1.115)$$

where, M_{de} = decimal equivalent of total material by volume

MM_{de} = decimal equivalent of magnetic material by volume referenced to total material

μ_{MM} = permeability of magnetic material

NMM_{de} = decimal equivalent of non-magnetic material by volume referenced to total material

A_{de} = decimal equivalent of air by volume = $1-M_{de}$

Eq. (1.115) for the above example becomes:

$$\mu_r = 0.80\text{x}0.60\text{x}1000 = 480$$

Thus, the low-frequency magnetic shielding effectiveness is:

$$SE_{dB} = 20 \log_{10}\left(1 + \frac{480\text{x}0.020}{2\text{x}0.50}\right) = 20.5 \text{ dB}$$

1.4.6 PERFORMANCE DEGRADATION

All preceding discussions assume that the shielding material is both homogeneous and large in planar dimensions such that neither leakages nor edge effects take place. This would otherwise result in a shield whose performance is compromised. The shielding effectiveness expressed by Eq. (1.64) may then be conceptually redefined:

$$SE'_{dB} = A_{dB}+R_{dB}+B_{dB} - \text{Leakage Effects} - \text{Standing Wave Effects} \qquad (1.116)$$

where, A_{dB} = absorption loss, Eq. (1.65) or (1.72)

R_{dB} = reflection loss, Eq. (1.66) or (1.82)

B_{dB} = re-reflection correction, Eq. (1.67) or (1.91)

Leakage effects in Eq. (1.116) may be identified as due to one or more of the following situations which exist in any practical, real-life shield:

- Seams
  - Mating members
  - Screws
  - Crimps
  - Welds
  - Braizes
  - Silver solder
  - Soft solder
- Doors
- Cover Plates
  - Access panels
  - Drawers
- Vents
  - Ventilation
  - Air conditioning
  - Heating
- Holes and apertures for:
  - Connectors
  - Fuses
  - Power lines
  - Signal and control lines
  - Fiber-Optic lines
  - Potentiometers
  - Viewing Windows
  - Status indicators
  - Push buttons
  - On/Off switches
  - Control switches
  - Convection cooling
- Non-Homogeneous Areas
  - Screens
  - Shield braids
  - Meshes
  - Thin areas

Degradation due to standing waves in Eq. (1.116) involves resonance effects at higher frequencies where enclosures act as microwave cavities. This results in areas or regions within a shielded enclosure which exhibit poorer performance, i.e., lower shielding effectiveness.

Eq. (1.116) is expressed in conceptual form rather than in explicit form. The method of combining effects due to shielding corresponding to leakages (L_{dB}) and the base metal shielding effectiveness, SE_{dB} (without these effects) is:

$$SE'_{dB} = -20 \log_{10}\left[\log_{10}^{-1}\left(-SE_{dB}/20\right) + \sum_{i=1}^{n} \log_{10}^{-1}\left(-L_{idB}/20\right)\right] \quad (1.117)$$

Note that Eq. (1.117) coherently adds all leakage effects together and coherently adds these results with SE_{dB}*. This is a worst case since the phases may range from nearly all in phase (coherent at low frequencies) to approaching random phase (incoherent at high frequencies)*. However, usually one or a few leakages predominate so that coherent worst-case combination is justified.

<u>Example</u>: Suppose that the shielding material, per se, of an enclosure results in a shielding effectiveness of 110 dB at some frequency, but the leakages separately yield shielding effectiveness of: (1) filter/

---

* If coherent addition did not take place (i.e., effects combine randomly or incoherently) the three "20s" in Eq. (1.117) would be replaced by three 10s.

connector panel = 101 dB, (2) vent leakage = 92 dB, (3) door leakage = 88 dB, and (4) shield panel seam leakage = 83 dB. Calculate the total leakage shielding and the combined shielding effectiveness of the enclosure.

Application of the second half of Eq. (1.117) yields a total leakage shielding effectiveness of 76.8 dB. Applying all of Eq. (1.117) gives SE'_{dB} = 76.6 dB. Thus, in this example, it does not particularly matter what the basic enclosure material performance is as long as it is about 10 dB better than the lowest or least leakage point. This may be illustrated by applying Eq. (1.117) for several different values of SE_{dB} in the above example having a total leakage shielding effectiveness of 76.8 dB:

Basic Enclosure Metal SE_{dB}	Shielding for Total Leakages L_{dB}	Combined Shielding SE'_{dB}
60 dB	76.8 dB	58.8 dB
70 dB	76.8 dB	66.7 dB
80 dB	76.8 dB	72.2 dB
90 dB	76.8 dB	75.1 dB
100 dB	76.8 dB	76.2 dB
110 dB	76.8 dB	76.6 dB
120 dB	76.8 dB	76.7 dB
130 dB	76.8 dB	76.8 dB
$\geq$ 150 dB	76.8 dB	76.8 dB

The above serves to demonstrate the importance of controlling leakages in shielded enclosures. Usually, but not always, a leakage results in a poorer performance than that of the base metal (one exception is for low-frequency magnetic fields where the phase shift through the leakage can add out-of-phase with the penetration of the base shield metal). The topic of leakages and the techniques for controlling them in shielded housing designs is discussed in Chaps. 11 and 12 of Vol. 3 of the EMI/EMC Handbook Series.

As it develops then, the shielding effectiveness of a useful six-sided homogeneous box may be regarded as an upper limit which approaches the analytical relations and design graphs of this handbook. Where overall shielding effectiveness is relatively low, say less than about 40 dB, there usually exists good correlation between theory and practice, except possibly for low-frequency magnetic-field sources. When SE_{dB} is, perhaps, between 40 and 100 dB, the correlation between theory and practice may range from good to poor depending upon (1) the degree of departure from an homogeneous enclosure, (2) the frequency, and (3) the degree to which the leakage points or areas have been shored up. Where correlation is not good, the measured values are almost always below theoretical values. For theoretical SE_{dB} values in excess of about 100 dB, correlation between theory and practice is generally poor since there are so many leakage joints in practical enclosures.

For magnetic materials ($\mu_r > 1$), the relations in this handbook apply for the stipulated permeability μ_r condition only; μ_r varies with both magnetic-field strength (or flux density) and frequency, especially above several kHz (see Sec. 2.1.4). Above about 1 MHz the relative permeability of most magnetic materials has already become or approaches unity. Thus, magnetic materials should not be used unless low-frequency shielding of magnetic fields is required.

1.4.6.1 MORE ABOUT LEAKAGES

Metals, which on an incremental surface-area basis do not exhibit nearly identical conductivity or surface impedance and the same thickness throughout, may be regarded as non-homogeneous metals (see Chap. 2). Examples of these include:

(1) The situations listed in the previous section
(2) Any practical and useful six-sided enclosure, whether at the circuit, equipment, rack, console, or room-size level
(3) Conductive paints, flame-spray deposits, very-thin depositions, metalized textiles, screens, wire mesh, and the like.

From the above list, it would appear that no useful shielded enclosure qualifies as a homogeneous metal for which the relations in this handbook can be used. In a strict sense, this is true since at the very least any enclosure, to be useful, must have wires (or fiber optics) entering and/or leaving the enclosure. Not only do these wires (except fiber optics) constitute either a radiating or pickup *antenna-like* source, they require holes, slits or apertures in the enclosure in order to be admitted. Again Chaps. 11 and 12 of Vol. 3 of the EMI/EMC Handbook Series discuss these situations in length.

For the relations of Sec. 1.4 of this handbook to apply, the metal should be homogeneous and infinite in the planar dimensions. As a practical matter, as long as either (1) the metal sheet planar dimensions are much greater than the distance between an emission source and the shield or (2) a six-sided enclosed box or parallelepiped is used, then the shielding effectiveness relations become an upper limit of performance.

Lack of enclosure homogeneity gives rise to low-frequency resonance effects which result from unequal phase shifts in wave transmission to a receiving point via two or more paths: one through the base shielding material and another or others through the seam-leakage paths. The resonant frequencies correspond to those situations for which the path lengths through the metal* are odd multiples of a half wavelength

* Path length differences outside the metal result in negligible phase differences in air at low frequencies and are neglected.

greater than through the leakage-point paths (usually air gaps). Thus, from Eq. (1.71) for the path length difference to be $\beta t = \pi$ (i.e., 180° path delay), the lowest resonant frequency is obtained, viz:

$$e^{j\beta t} = e^{j\pi} = e^{jt\sqrt{\pi f \mu \sigma}} \quad (1.118)$$

or,

$$f_r = \frac{\pi}{t^2 \mu \sigma} = \frac{6.68 \times 10^7}{t^2_{mils} \mu_r \sigma_r} \text{ English units} \quad (1.119)$$

$$= \frac{4.31 \times 10^4}{t^2_{mm} \mu_r \sigma_r} \text{ metric units} \quad (1.120)$$

Example:

Calculate the magnetic-field resonant frequencies of a shielding box fabricated from 1/32" aluminum and 1/32" steel.

Here, t_{mils} = 31.25 mils (1.32" = 0.794mm) and σ_r = 0.63 and μ_r = 1 for aluminum, and σ_r = 0.17 and μ_r = 1000 for steel. From Eq. (1.119), the resonant frequencies are: 109 kHz for aluminum and 402 Hz for steel.

Below resonance, the shielding effectiveness of a leaky box is fairly well defined by the shielding effectiveness relations of the base material. At and very near low-frequency resonance, the shielding effectiveness is greater, and above resonance, SE_{dB} is less than that defined by the performance of the base material. A further discussion of these topics is beyond the scope of this handbook because the results are seam-type and geometry related. For further information, the reader is referred to the work of Schultz, Cooley and others mentioned in the references in Chap. 4 of this handbook and again, in Chaps. 11 and 12 of Vol. 3 of the EMI/EMC Handbook Series.

1.4.6.2 STANDING WAVES

As a result of either penetrations or leakages, a six-sided shielded box or cabinet will also result in internal reflections. It can act as a cavity with standing waves at higher frequencies. Resonant frequencies*, f_{MHz}, of such an enclosure exist at:

$$f_{MHz} = 150 \sqrt{\left(\frac{k}{\ell}\right)^2 + \left(\frac{m}{h}\right)^2 + \left(\frac{n}{w}\right)^2} \text{ MHz} \quad (1.121)$$

where, ℓ = length of box or enclosure in meters

* The microwave cavity resonance effect here is not to be confused with the low-frequency resonance discussed in the previous section, Sec. 1.4.6.1. The two effects are entirely different phenomena.

h = height of box or enclosure in meters
w = width or depth of box or enclosure in meters
k,m,n = a positive integer 0, 1, 2, 3....but not more than one of k,m, or n can be zero at the same time to support any resonant mode.

From Eq. (1.121), if the box or enclosure dimensions are all comparable, i.e., are approximately equal, then the resonant frequencies, f_{MHz}, corresponding to the lowest order TE_{011}, TE_{101}, or TE_{110} modes are:

$$f'_{MHz} \simeq 212/\ell \simeq 212/h \simeq 212/w \qquad (1.122)$$

Thus, the shielding effectiveness of boxes or enclosures will be expected to exhibit wide variations on the inside, above those frequencies corresponding to approximately 212/d, where d is the average dimension in meters. These variations come about as a result of the tangential electric-field intensity becoming the least at the inside of the metal enclosure and rising to greater values within.

Illustrating the above, the lowest resonant-frequency effects associated with (1) a one-foot cubed box $(.305\ m)^3$; (2) a five-foot (1.52 m), 19 inch wide rack (measures 22' ID or 0.56 m) of 30" (0.76 m) depth; and (3) an 8 ft. (2.44 m) x 12 ft. (3.66 m) x 20 ft. (6.10 m) shielded enclosure are, respectively: 696 MHz, 220 MHz, and 47.8 MHz.

The shielding effectiveness relations in this handbook do not account for any standing-wave effects. In general, the standing waves inside of a six-sided box or enclosure will result in fields which vary from twice that through the basic enclosure material to very low values at the metal interfaces. Thus, a maximum 6-dB degradation in shielding effectiveness can be expected from enclosure resonances, per se.

CHAPTER 2

SHIELDING MATERIALS AND TESTING

Among other things, the previous chapter indicated that for the equations of Sec. 1.4 to apply, the shielding material characteristics must be homogeneous and independent of frequency and field strength or other saturating effects (i.e., they are linear). These equations apply, with restrictions, to pseudo-homogeneous metals such as conductive paints, coatings, depositions, flame-spray processes and the like. Except at lower frequencies, they do not apply for small-aperture materials such as screens, meshes, and metalized textiles which follow an entirely different law. These and other related topics on shielding materials are presented in this chapter.

2.1 SHIELDING MATERIALS

This section summarizes the properties of homogeneous metals, pseudo-homogeneous metals and small-aperture metals.

2.1.1 Homogeneous Metals

Many of the metals and their associated conductivity and permeability which the designer might expect to encounter are listed in Tab. 2.1. The fourth column, entitled, *Product* $\sqrt{\sigma_r \mu_r}$ is a ranking of the latent absorption loss of metals relative to copper. For non-magnetic materials, except silver, all $\sqrt{\sigma_r \mu_r}$ values relative to copper are less than 1.0. Thus, non-magnetic materials provide relatively poor absorption loss. All magnetic materials, on the other hand, have relative absorption loss values exceeding two and are relatively good absorbers of energy at low frequencies compared to non-magnetic materials. On the other hand, since the relative permeability degrades with frequency, magnetic materials offer absorption losses less than most non-magnetic materials at higher frequencies (above approximately 100 kHz).

For thin-metal thickness (thicknesses for which $t/\delta \ll 1$, i.e., the metal thickness is a small percentage of the skin depth) the absorption loss is negligible (see Eqs. (1.98)-(1.100)) and choice of metal, whether magnetic or non-magnetiic, for shielding effectiveness based on absorption loss alone is unimportant. However, when the metal is not

Table 2.1 - Relative Conductivity and Permeability of Metals

(1) Metal	(2) Relative Conduct. σ_r	(3) Relative Permbly. μ_r@ $\leq$ 10kHz	(4) Product $\sqrt{\sigma_r \mu_r}$ $A = k_1\sqrt{\sigma_r \mu_r}$	(5) Quotient $\sqrt{\sigma_r/\mu_r}$ $R = k\sqrt{\sigma_r/\mu_r}$	(6) Relative Reflection R_{dB}
1. Silver	1.064	1	1.03 dB	1.3	+0.3 dB
2. Copper (Solid)	1.00	1	1	1	0
3. Copper (Flame Spray)	0.10	1	0.32	0.32	-10.0
4. Gold	0.70	1	0.88	0.88	-1.1
5. Chromium	0.664	1	0.81	0.81	-1.8
6. Aluminum (Soft)	0.63	1	0.78	0.78	-2.1
7. Aluminum (Tempered)	0.40	1	0.63	0.63	-4.0
8. Aluminum (Household Foil, 1 mil)	0.53	1	0.73	0.73	-2.8
9. Aluminum (Household Foil, 1 mil)	0.61	1	0.78	0.78	-2.1
10. Aluminum (Flame Spray)	0.036	1	0.19	0.19	-14.4
11. Brass (91% Cu, 9% Zn)	0.47	1	0.69	0.69	-3.3
12. Brass (66% Cu, 34% Zn)	0.35	1	0.52	0.52	-5.7
13. Magnesium	0.38	1	0.61	0.61	-4.3
14. Zinc	0.305	1	0.57	0.57	-4.9
15. Tungsten	0.314	1	0.56	0.56	-5.0
16. Beryllium	0.33	1	0.53	0.53	-5.5
17. Cadmium	0.232	1	0.48	0.48	-6.3
18. Platinum	0.17	1	0.42	0.42	-7.6
19. Tin	0.151	1	0.39	0.39	-8.2
20. Tantalum	0.12	1	0.33	0.33	-9.6
21. Lead	0.079	1	0.28	0.28	-11.0
22. Monel (67 Ni, 30 Cu, 2 Fe, 1 Mn)	0.041	1	0.20	0.20	-13.9
23. Manganese	0.039	1	0.20	0.20	-14.1
24. Titanium	.036	1	0.19	0.19	-14.4
25. Mercury (Liquid)	0.018	1	0.13	0.134	-17.4
26. Nichrome (65 Ni, 12 Cr, 23 Fe)	0.0012	1	0.035	0.035	-29.2
27. Supermalloy	0.023	100,000	53.7	0.0005	-65.4
28. 78 Permalloy	0.108	8,000	29.4	0.0037	-48.7
29. Purified Iron	0.17	5,000	29.2	0.0058	-44.7
30. Conetic AA	0.031	20,000	28.7	0.0011	-58.8
31. 4-79 Permalloy	0.0314	20,000	25.1	0.0013	-58.0
32. Mumetal	0.0289	20,000	24.0	0.0012	-58.4
33. Permendur (50 Cu, 1-2V, & Fe)	0.247	800	14.1	0.018	-35.1
34. Hypernick	0.0345	4,500	12.5	0.0028	-51.1
35. 45 Permalloy (1200 Anneal)	0.0384	4,000	12.4	0.0031	-50.2
36. 45 Permalloy (1050 Anneal)	0.0384	2,500	9.80	0.0039	-48.1
37. Hot-Rolled Silicon Steel	0.0384	1,500	7.59	0.0051	-45.9
38. Sinimax	0.0192	3,000	7.59	0.0025	-51.9
39. 4% Silicon Iron (Grain Oriented)	0.037	1,500	7.45	0.0050	-46.1
40. 4% Silicon Iron	0.029	500	3.81	0.0076	-42.4
41. 16 Alfenol	0.0113	4,500	7.13	0.0016	-56.0
42. Hiperco	0.069	650	6.70	0.010	-39.7
43. Monimax	0.0216	2,000	6.57	0.0033	-49.7
44. 50% Nickel Iron	0.0384	1,000	6.20	0.0062	-44.2
45. 45-25 Perminvar	0.091	400	6.03	0.015	-36.4
46. Commercial Iron (0.2% Impure)	0.17	200	5.83	0.29	-30.7
47. Cold-Rolled Steel	0.17	180	5.53	0.031	-30.2
48. Nickel	0.23	100	4.70	0.047	-26.6
49. Stainless Steel (1Cu,18Cr,8Ni,&Fe)	0.02	200	2.00	0.010	-40.0
50. Rhometal (36 Ni)	.019	1,000	4.36	0.0044	-47.2
51. Netic 53-6	.172	300	7.18	0.024	-32.4

thin (when t/δ is *not* << 1), the absorption loss becomes significant. Apart from cost and other considerations, Cols. 2 and 3 are useful in choosing a metal for shielding or simply in extracting the applicable σ_r and μ_r values for use elsewhere in this handbook.

The fifth column of Tab. 2.1 entitled *Quotient* $\sqrt{\sigma_r/\mu_r}$ is a ranking of the latent reflection loss of metals relative to copper. Expressed in dB form in Col. 6, the non-magnetic materials outperform nearly all magnetic materials for reflection-loss performance. For thin-metal thickness (thicknesses for which $t/\delta \ll 1$), the reflection loss is the only important term for overall shielding effectiveness. For this condition then, the designer would select a non-magnetic material of higher conductivity.

For conductive coatings on glass or plastic substrates used for R-F shields in which optical viewing is required, where t is measured in nm (a tiny fraction of δ), the more stable (lower on the electrochemical series), higher-conductive metals would be selected, such as gold. The shielding effectiveness for gold corresponding to $t/\delta \ll 1$ is plotted in Figs. 2.17 and 2.18.

When dealing with magnetic shielding materials, the question arises regarding the numerical value of relative permeability to be employed in the calculation of shielding effectiveness discussed in Sec. 1.4. The values of relative permeability, μ_r, given in the various literature are often used directly, as a fixed, material parameter which is non-varying with external conditions. Unfortunately, materials that fall into the classification of *magnetic,* i.e., $\mu_r > 1$, are subject to, and respond to material external environmental effects, not the least of which is the ambient magnetic environment. The evaluation of SE_{dB} in a realistic manner depends on the utilization of the correct value of μ_r. Factors needing consideration in the development of accurate permeability values are discussed below.

Fig. 2.1 shows the magnetization curve and saturation-to-saturation hysteresis curve of a *typical* material. In this figure various characteristics of the material are defined. H_c is the coercive force or the value of applied reverse magnetic-field strength required to drive the magnetic induction (flux density) to zero. B_r is the residual magnetism or value of magnetic induction retained after a saturation value of induction, B_m, has been achieved by application of a magnetic-field strength, H_m, which is then reduced to zero.

Relative permeability, μ_r, is defined as the ratio of magnetic induction B to applied magnetic-field strength H. Examination of the fig. shows the definition of three different relative permeabilities, μ_r: μ_o, the initial permeability; μ_{max}, the maximum permeabilities; and μ_Δ the incremental permeability. These three values of permeability can vary by more than an order of magnitude (or more than 10 dB of SE_H for reflection loss and by any amount for absorption loss). Since

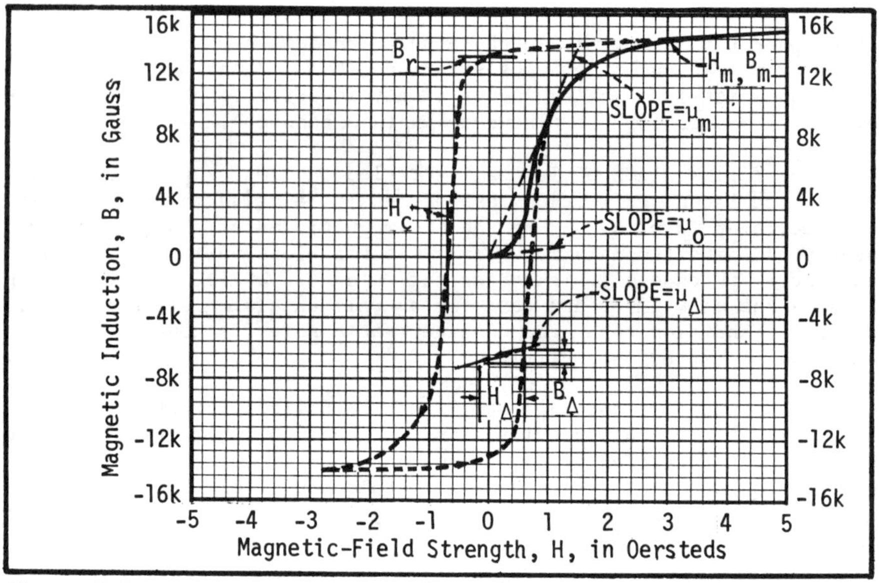

Figure 2.1 - Magnetization Curve (solid) and Hysteresis Loop (dotted). Some Important Magnetic Quantities are Illustrated.

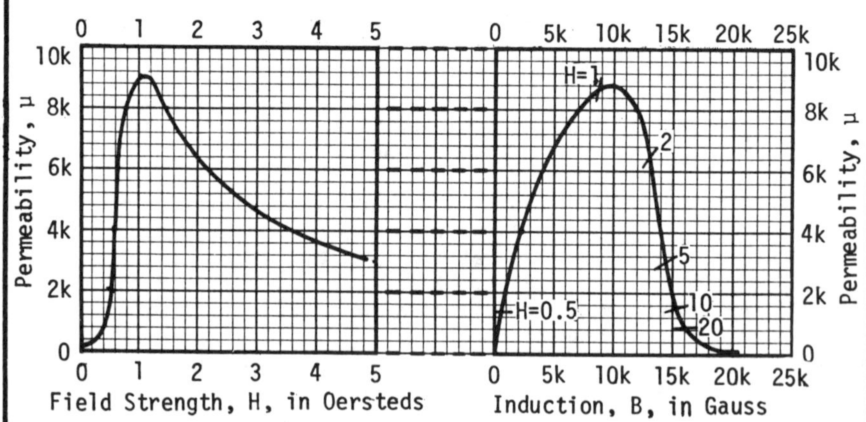

Courtesy of Bozorth, R.M., Ferromagnetism, D. Van Nostrand Co., Inc., New York

Figure 2.2 - Permeability Curves of Iron, with μ Plotted Against H and B. I and B-H Are Also Used As Abscissae.

magnetic SE_{dB} values are usually low at low frequencies, this can result in a significant error.

The incremental permeability, μ_Δ, is of particular interest in this discussion. As shown in the figure as a minor loop on the hysteresis curve it shows the effect of a low-level signal applied in addition to the hysteresis loop generation signal, at the single point shown, in opposition to the signal. This represents the situation when a magnetic field strikes a permeable shield.

Fig. 2.2 shows the relative permeability, μ, as a function of applied magnetic-field strength and induction. This clearly portrays the dependence of permeability upon the magnetic environment, a factor which must be considered in evaluating μ for the computation of SE_{dB}.

Removing the hysteresis curve, while retaining the presence of the externally applied magnetic signal results in the situation shown in Fig. 2.3 where the small loops generated by applied fields are shown on the magnetization curve. Note that μ_Δ will vary with applied field, in that the B-H slope varies, which defines μ_Δ.

The situation experienced by a shield material in real life is, of course, even more complex. Shields are only used in the presence of alternating fields. Thus, one typical situation would be a shield existing in the earth's magnetic field ($\sim$0.5 Oersted*) of D-C bias, with typically a 60-Hz H-field superimposed causing a hysteresis effect and (at a minimum) the signal to be shielded. As shown in Fig. 2.4, this situation allows the estimation of an effective permeability, μ_e, to be used for computation of SE_{dB}:

$$\mu_e \simeq \mu_\Delta \simeq 2\mu_o = 0.1\mu_m$$

In the more general situation, several effects take place including eddy current losses which are a function of (1) magnetic-field strength, (2) frequency, (3) distances to the metal barrier, (4) metal thickness, (5) metal permeability and resistivity. Thus, it is concluded that unless all conditions are known, it is impossible to define an equivalent μ_r and to directly use the μ_r values in Tab. 2.1 and the equations in Sec. 1.4 for magnetic materials.

Fig. 2.5 shows $\mu_r \rightarrow 1$ above a few hundred kHz for an undefined magnetic-field strength, whereas Fig. 2.6 shows the behavior of μ_r vs. magnetic-flux density** at 60 Hz. About all one might conclude from this is that below saturation and at very low frequencies the permeability can increase up to about an order of magnitude above its rated low-

* Magnetic-field strength of 1 Oersted = $10^3/4\pi$ = 79.58 ampere-turns/meter.

** 1 gauss = 10^{-4} Tesla = 160 dBpT.

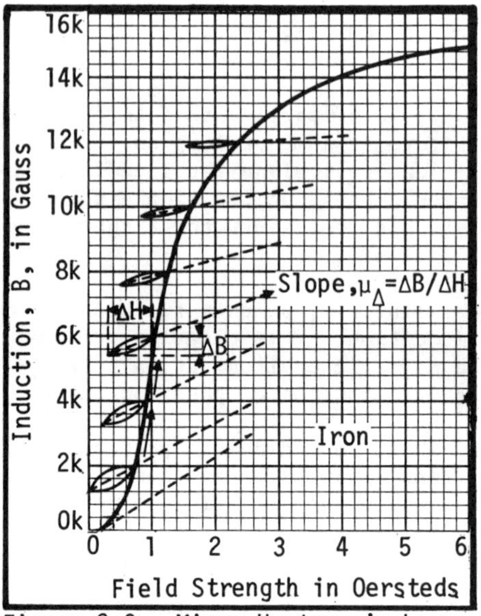

Figure 2.3 - Minor Hysteresis Loops Shown on Magnetization Curve

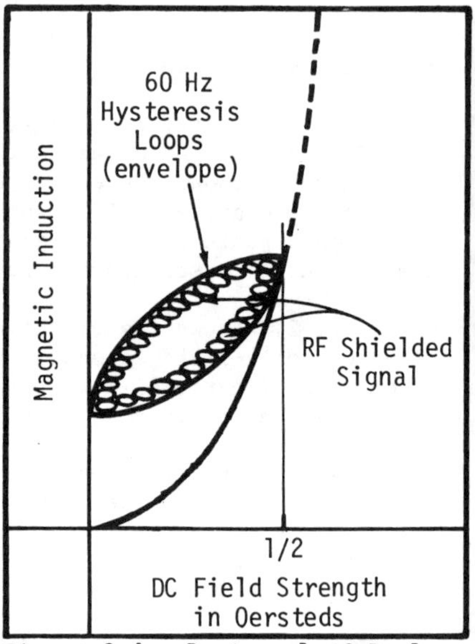

Figure 2.4 - Portrayal of Real World Situation

level values listed in Tab. 2.1. At and above medium frequencies (300 kHz - 3 MHz), $\mu_r \rightarrow 1$ for all conditions. Thus, the shielding effectiveness curves presented in the appendices apply only for the stated values of μ_r used.

When the maximum magnetic-flux density may be exceeded, a double shield containing either a non-magnetic metal or a higher saturation magnetic metal should face the more hostile magnetic-field source. For hostile emissions coming from outside a box or cable shield, this means that the first layer of protective metal should face the outside, whereas if the emissions originate within, the protective metal should face the inside.

2.1.1.2 AVAILABILITY AND APPLICATIONS

Several of the metals in Tab. 2.1 are available off-the-shelf in sheet stock form from thicknesses of about 1/64th inch (0.4 mm) or less to about 1/8th inch (3.2 mm) or more. Metals having thicknesses less than 1/64th inch are sometimes regarded as foils. Many of the high-permeability metals come in foil thicknesses ranging from about 1 mil (25.4 μm) to 10 mils (254 μm). They are usually available in both sheet and tape form. The foil stock is also available in the form of adhesive-backed foil in roll lengths typically up to 100 ft. (30.5 m).

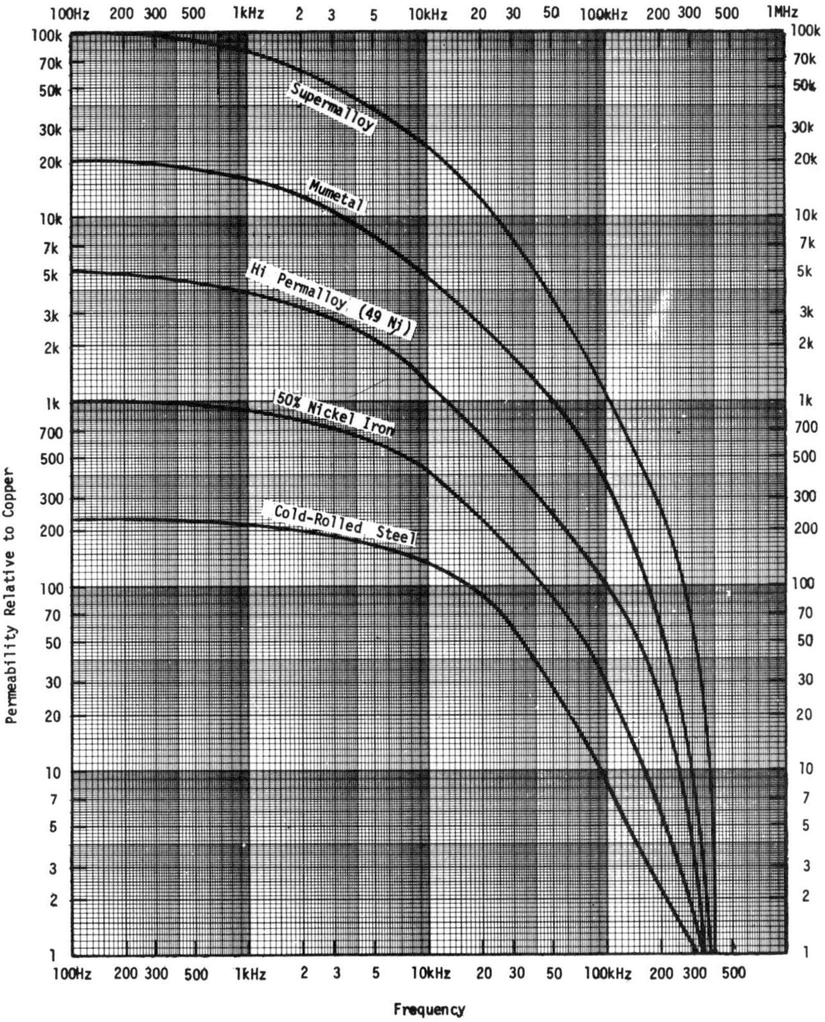

Figure 2.5 - Relative Permeability vs. Frequency

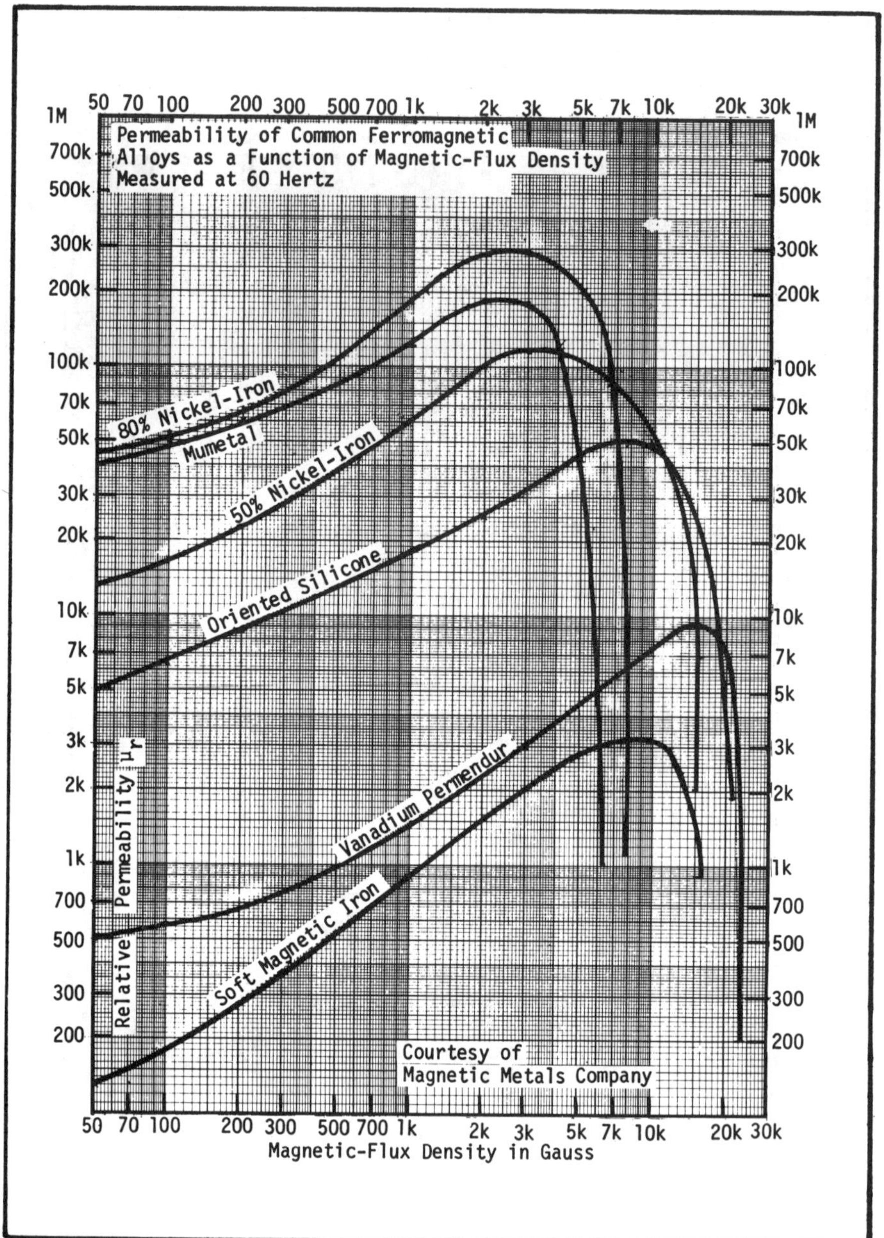

Figure 2.6 - Relative Permeability vs. Magnetic-Flux Density

The thinner non-magnetic foils, whose thickness is of the order of a few mils, are widely used for R-F shielding. While there are many stories spread around about how *Reynolds Wrap* saved the day, the appendices clearly show that one mil of household aluminum foil can produce shielding effectiveness to plane waves and electric fields in excess of 80 dB below 1 GHz (see Sec. 2.2). Its performance to low-frequency magnetic fields, however, is very poor, i.e., is nearly transparent.

Metal foils are used in a number of ways. One rather interesting application involves metal-foil wallpaper (MFWP) for converting an entire room into a limited shielded enclosure. It exhibits considerably less shielding effectiveness than purchased shielded rooms, but also at a much lower cost. These materials have been used on occasion to construct shielded chambers of thousands of square feet (hundreds of square meters) in surface area down to small equipment enclosures. MFWP must be used in conjunction with other materials such as pressure-sensitive metal-foil tape, conductive adhesives, or conductive epoxy or caulking compounds. They also require taking a number of measures to reinstate lost shielding integrity at doors, windows and the like.

MFWP usually comes in thicknesses of two to three mils (51 to 76 microns) and is made of either aluminum or copper foil or special stainless steel foil of relatively high conductivity and high permeability. Representative shielding effectiveness is 25-40 dB for magnetic fields at 200 kHz, 80-100 dB for electric fields from 200 kHz to 10 MHz, and 60-80 dB for plane waves above 400 MHz (see Sec. 2.2) when measured in accordance with MIL-STD-285 (see Sec. 2.3).

Sometimes thin foils of the order of one mil are bonded to (metalized on) a plastic base such as 5-10 mil mylar. They can then be used as an air-inflatable structure which performs as a shielded enclosure. Such applications are discussed in Sec. 3.1.3 of Vol. 2 of the EMI/EMC Handbook Series.

Some of the magnetic sheet metal and/or foil manufacturers who make the basic stock supplied to the EMI/EMC and related communities include:

Magnetic Shield Division
Perfection Mica Company
740 Thomas Drive
Bensenville, Ill. 60106
Phone: (312) 766-7800
TWX: (910) 256-4815

The Inter-Technical Group, Inc.
P.O. Box 23
Irvington, New York 10533
Phone: (914) 591-8822
TWX: (710) 564-0802

AD-VANCE Magnetics, Inc.
226 East 7th Street
Rochester, Indiana 46975
Phone: (219) 223-3158

Eagle Magnetic Company, Inc.
P.O. Box 24283
Indianapolis, Indiana 46224
Phone: (317) 297-1030

Micrite Company
21531 Strathern Street
Canoga Park, Calif. 91304
Phone: (213) 348-1610

James Millen Mfg. Co., Inc.
150 Exchange Street
Malden, Mass. 02148
Phone: (617) 324-4108

2.1.2 PSEUDO-HOMOGENEOUS METALS

Metals which lack homogeneity but which do not *intentionally* have holes, slits, or other apertures, whether small or large, are called pseudo-homogeneous metals (PHM) in this handbook. Examples of PHM include conductive paints and coatings and the flame-spray process of metalizing an insulator. PHM may have *thin areas,* because lack of adequate quality control of the very process, per se, does not lend itself to homogeneity. Consequently, PHM may result in theoretical shielding effectiveness which may compare anywhere from good to poor with respect to measured results.

The electrical properties of pseudo-homogeneous, conductive coatings are usually measured in units of either surface resistance per unit thickness, or in units of volume resistivity, absolute conductivity, or occasionally conductivity relative to copper. The surface resistance, R_s, is measured in ohms/sq.* and is related to volume resistivity,** ρ:

$$R_{sdc} = 100\rho_{\Omega\text{-m}}/t_{cm} \quad \text{ohms, metric system} \tag{2.1}$$

$$= 39370\rho_{\Omega\text{-m}}/t_{mils} \quad \text{ohms, English system} \tag{2.2}$$

where, t = thickness of surface coating in indicated units

The volume resistivity is related to conductivity, σ, and surface resistance in the following manner at DC (1.108A):

$$\sigma_{mhos/m} = 1/\rho_{\Omega\text{-m}} = 100/R_{sdc}t_{cm} \quad \text{metric system} \tag{2.3}$$

$$= 39370/R_{sdc}t_{mils} \quad \text{English system} \tag{2.4}$$

Finally, the conductivity relative to copper (σ_r for copper = 1; σ_{cu} = 5.8×10^7 mhos/m) is:

---

* The term ohms/sq. refers to the D-C resistance of a square cross section having equal dimensions on each side.

** The author has observed a more or less equal split in the literature between those reporting resistivity in units of ohms-cm and ohms-m. Thus, he has arbitrarily selected the MKS system of ohms-m.

$$\sigma_r = \sigma_{metal}/\sigma_{cu} = 1.72 \times 10^{-8}/\rho_{\Omega\text{-}m} \quad (2.5)$$

$$= 1.72 \times 10^{-6}/R_s t_{cm}, \quad \text{metric system} \quad (2.6)$$

$$= 6.79 \times 10^{-4}/R_s t_{mils}, \quad \text{English system} \quad (2.7)$$

Eqs. (2.1) and (2.2) are plotted in Fig. 2.7 to yield surface resistance vs volume resistivity with metalized coating thickness as a parameter. Eqs. (2.6) and (2.7) are plotted in Fig. 2.8 to present coating conductivity relative to copper vs surface resistance with thickness as a parameter. This figure is especially useful since it allows developing an equivalent relative conductivity, σ_r, which can be used in the many expressions of shielding effectiveness throughout this handbook.

2.1.2.1 CONDUCTIVE PAINTS

Conductive paints (also called conductive coatings) are made by Emerson and Cuming (E&C), Technical Wire Products (Tecknit), Acheson Colloids, and others (see references). They are usually a lacquer, elastomeric, silicone resin, vinyl, acrylic, or latex base and require careful surface preparation. Some conductive paints require an overcoat for protection. To assure good electrical and mechanical reliability, surface preparation of plastics, woods, ceramics, and other base materials requires removing all greases, waxes, oils, dirt, mold releases and foreign matter until the surface is water-break free. Most conductive paints may be applied by one or more of the following processes: dipping, spraying, silk screening, roll coating, or brushing. Several conductive paints are also available in aerosol spray cans.

Silver lacquer paint (e.g., E&C Eccocoat CC-2) is a highly conductive silver particulate and organic resin formulation. In most cases, a single spray coat on a reasonably non-porous surface using air drying is adequate to produce a surface resistivity of about 0.1Ω/sq. for a 1 mil (25.4 μm) coating thickness. This corresponds to a volume resistivity of 0.04 ohm-m (σ = 25 mhos/m, or σ_r = 0.0042; see Figs. 2.7 and 2.8). Oven curing, if possible, gives improved conductivity.

The silver-filled elastomeric (e.g., E&C Eccocoat CC-4) is also highly conductive and a film will stretch over 100%. Thus, it is particularly useful where the substrate to which it is applied is subject to flexing or stretching. Air drying typically results in a surface resistance of about 50 mΩ/sq. for a 1-mil coating thickness. This corresponds to a σ_r = 0.014. Oven curing results in a surface resistance of about 1 mΩ/sq. or a σ_r = 0.68. It provides a resistance to salt spray and has an operating temperature range of -65°F (-54°C) to 250°F (121°C). Extended use at temperatures of 250° is excellent.

The silver-base, silicone (e.g., E&C Eccocoat CC-10) surface coating exhibits both high conductivity and high operational temperature

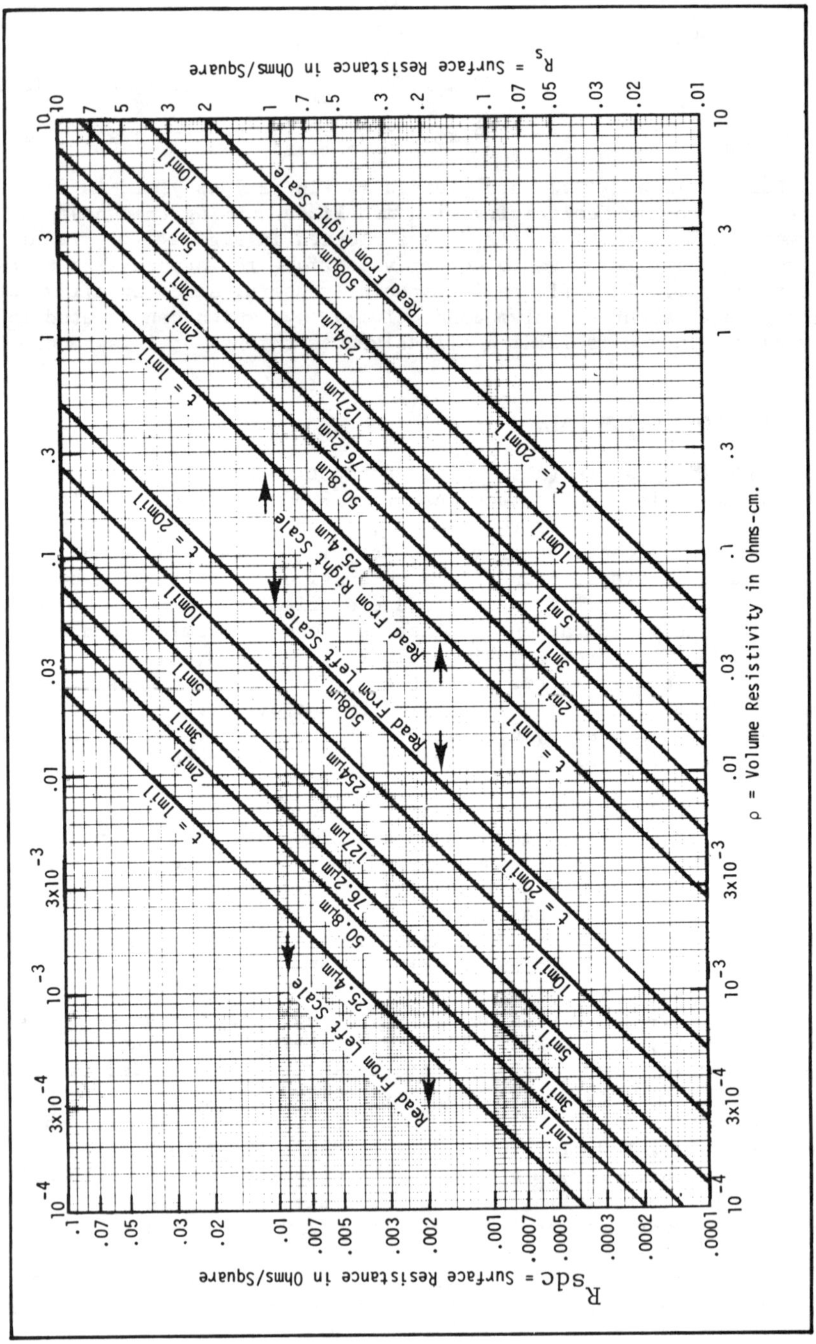

Figure 2.7 - Surface Resistance of Copper vs. Volume Resistivity for Various Metal Thicknesses

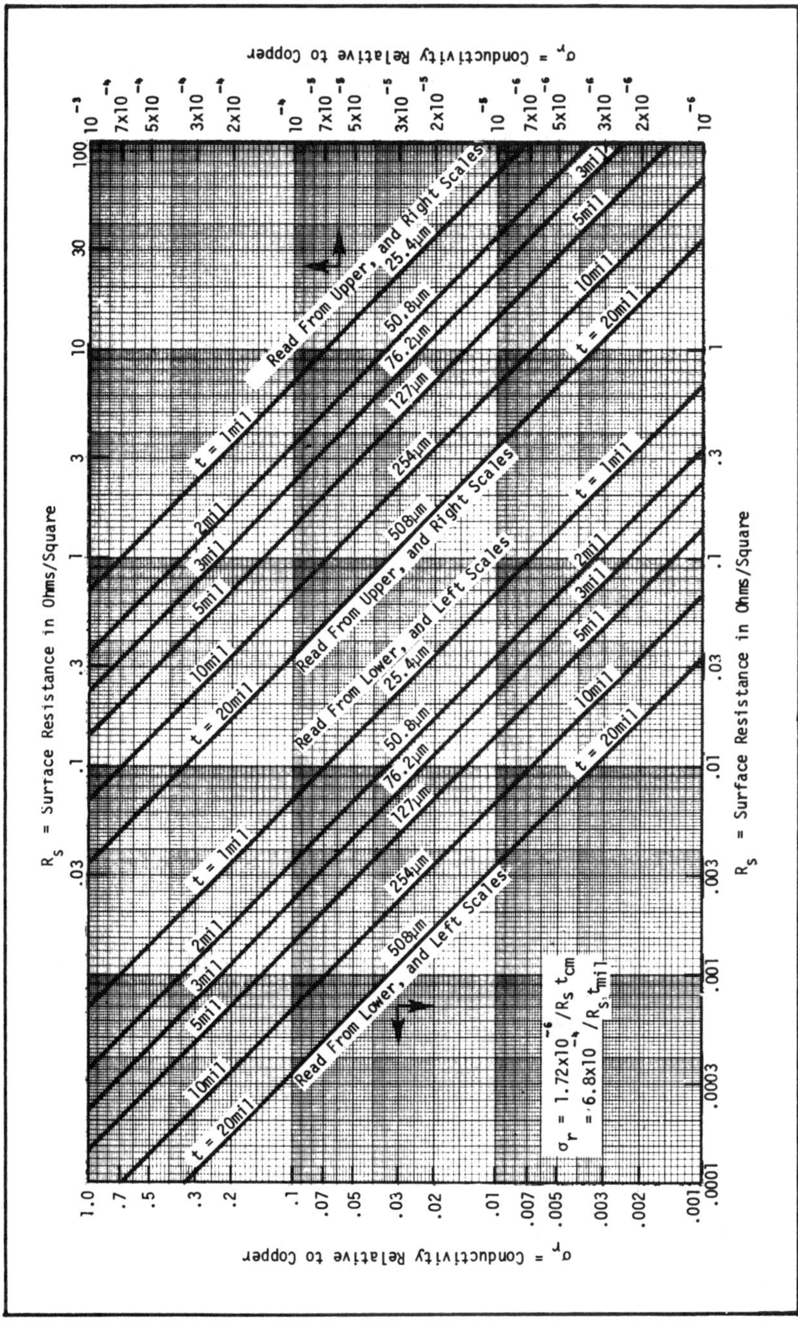

Figure 2.8 - Conductivity Relative to Copper vs. Surface Resistances

properties. The surface resistance is 40 mΩ/sq. for a 1-mil coating thickness (σ_r = 0.39). It can be used continuously at a temperature of 600°F (315°C). When applied to a substrate which will withstand even higher temperatures, the surface coating can be fused to the substrate. Moisture and chemical resistance are good.

Paints consisting of a pure silver filler in a vinyl copolymer (e.g., Tecknit 72-00026) result in a conformal coating having surface resistances of about 10 mΩ/sq. for a 1-mil deposit. It will stretch, compress, and deflect with its substrate while maintaining basic properties. The operating temperature range is -65°F (-54°C) to 250°F (121°C).

The conductive acrylic coatings developed for the plastic industry (e.g., Tecknit 72-00025) exhibit a surface resistance of about 50 mΩ/sq. for a 1-mil deposit. The variation depends upon the percent solids used, the base material, the deposition depth, etc. In preparation for spraying, the surface is typically cleaned with either alcohol or naphtha. Should environmental, decorative, and/or dielectric insulation protection be required on conductive acrylic paint, the overcoat must be a non-solvent paint. Typical overcoats exhibit dielectric strengths of 300 V/mil (12 kV/mm) and have temperature ranges which match the conductive coating: -65°F (-54°C) to 200°F (93°C).

A number of different conductive surface coatings called *Electrodag* are made by Acheson Colloids. They range in surface resistance from 15 mΩ/sq/mil for more expensive highly reflective applications to less expensive 200 Ω/sq/mil for low shielding effectiveness use. A particular coating is selected based on the abrasion resistance; surface hardness; adhesion to plastics, glass, or ceramics; chemical and solvent resistance; and service temperature. While conventional paint-spraying is the usual application, brushing or dipping techniques may also be used. Figs. 2.9 and 2.10 illustrate typical applications.

Prices for the above conductive coatings vary over a wide range depending upon the quantity, method of application, and type of conductive paint selected. In smaller quantities, prices may range from about 10 cents per sq. ft. ($0.093m^2$) per mil thickness for higher surface resistances to $4.00 per sq. ft. per mil thickness for low surface resistances with special other properties.

The above conductive coating manufacturers may be reached at the following:

Acheson Colloids Company
Div. of Acheson Industries
Port Huron, Michigan 48560
Phone: (313) 984-5581
TWX: (810) 231-5265

Emerson & Cuming, Inc.
869 Washington Street
Canton, Mass. 02021
Phone: (617) 828-3300

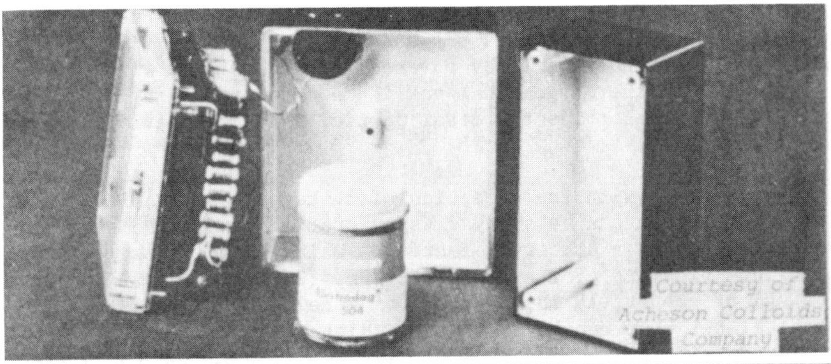

Figure 2.9 - Conductive Coating on Interior of Plastic Equipment Enclosure.

Figure 2.10 - Conductive Coatings for Both EMI and Static Bleed Functions.

Manufacturers' of Conductive Coatings (cont.):

Technical Wire Products, Inc.
129 Dermody Street
Cranford, New Jersey 07016
Phone: (201) 272-5500
TWX: (710) 996-5951

2.1.2.2 FLAME-SPRAY PROCESS

Flame spraying of metals is a process which applies in this handbook to metalizing a surface by flame vaporizing the basic metal which becomes deposited upon a surface. This process includes flame spraying, thermo spraying, and plasma spraying.

In flame spraying a desired wire rod of copper, aluminum, or other metal is fed into a special metalizing gun. This gun also accepts a special gas such as oxyacetylene or oxypropane and compressed air. One typical configuration is shown in Fig. 2.11. The wire rod is drawn through the gun and nozzle by powered feed rods. The wire is continually melted in the oxygen-fuel gas flame and atomized by a compressed air blast which carries the metal particles to the previously prepared surface.

Any metal (with the exception of tungsten) obtainable in a uniform wire or rod size can be sprayed using special guns. Flame spraying speeds depend on a number of factors. Using 15 psi (5 kg/cm^2) of acetylene pressure, for example, 12 lbs. (26 kg) of aluminum, 29 lbs. (64 kg) of copper, 18 lbs. (40 kg) of monel, or up to 16 lbs. (35 kg) of steel per hour can be deposited. Higher spraying speeds can be achieved using propane gas. Typically, metal wire and rod diameters from 18 B&S gauge (0.91 mm) to 3/16 inch diameter (4.76 mm) can be accommodated in the special spray guns.

Since the flame, (about 6000°F = 3300°C) which is located in the nozzle, does not dwell over any surface area, surface material such as plastics and fiberglass are not damaged. Air pockets in the deposition however, can develop and, depending upon the process control, the metalized surface may range from a sandpaper-type finish to a relatively smooth one. Chemical analysis indicates that oxides of the base metal deposited on the surface are low.

The flame-sprayed deposits of copper and aluminum listed in Tab. 2.1 exhibit conductivities relative to their own base metals of about 10% for copper to 6% for aluminum. This develops because most of the deposit is trapped air particles or tiny voids in the deposited surface, viz., the basic metal density is low. Consequently, it takes about 10 times for copper and 17 times for aluminum of the base metal thickness, per se, to yield the same conductivity. For thin deposits there is also the danger of many small area voids in the flame-spray process.

A variation of the flame-spray process, called *Electrospray* by Wall Colmonoy Corp., uses an electric-arc gun. Used for higher volume metalizing operations, it works faster and costs somewhat less to operate. The 7000°F (3900°C) temperature of the arc melts a wire rod faster and deposits particles having higher heat and greater fluidity. There is some danger, however, that plastic substrates could be deformed if the flame spray dwells too long on a surface.

The electric-arc spray rate is claimed to be three to five times faster than that of oxyacetylene equipment, and bond strength is greater as welded adhesion is added to the usual mechanical bond. This latter property, of course, cannot be used with plastic substrates which means that a minimum speed across a surface must be maintained. Good bonds can be achieved with less than ideal surface preparation. Since hotter particles mean denser coatings, coating oxide content is reduced compared to conventional oxyacetylene methods.

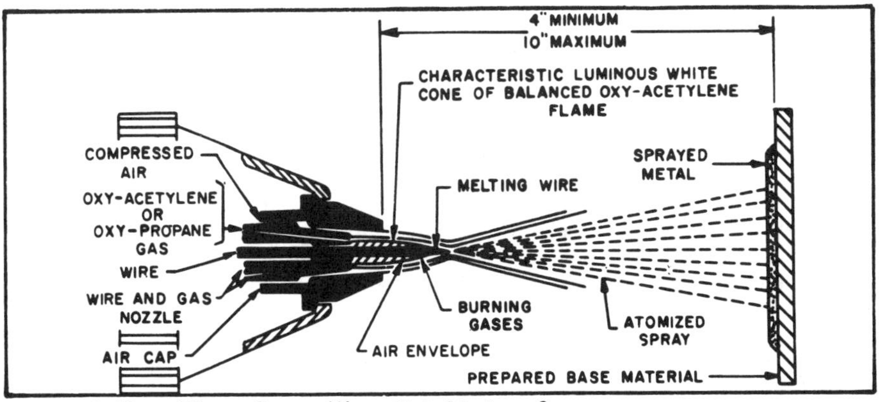

Figure 2.11 - Flame Spray Wire Metalizing Gun

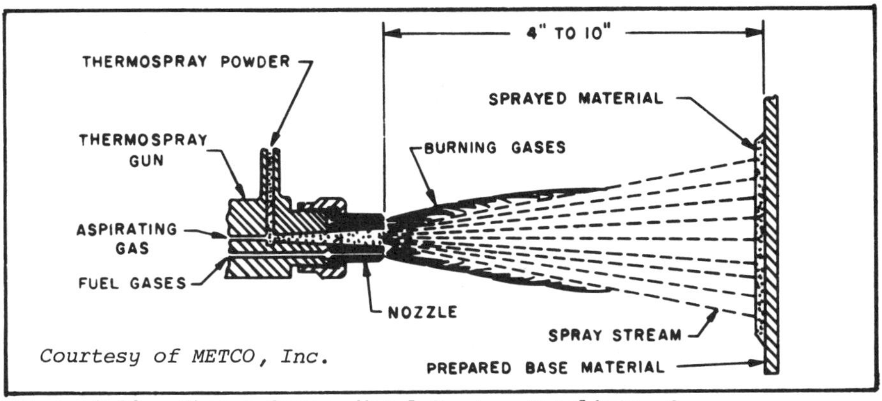

Figure 2.12 - Thermo Spray, Metal Powder Metalizing Gun

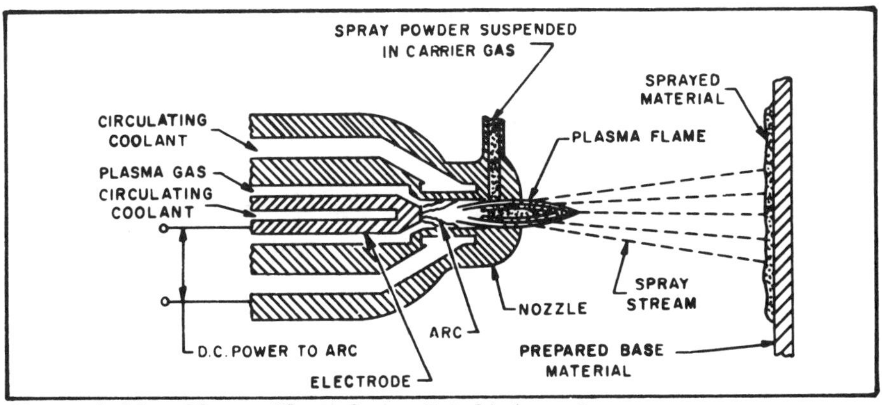

Figure 2.13 - Plasma Flame Spray Metalizing Gun

A different type of flame-spray process, called *ThermoSpray* by Metco, Inc. (see Fig. 2.12) accepts the metal to be sprayed in powder form. The powdered aluminum, copper, or other metal is held in a hopper atop the gun and gravity fed into the gun where it is picked up by oxyacetylene or hydrogen gas mixtures and carried to the gun nozzle. Here, it is melted almost instantly due to the thermal efficiency of the gun and carried to the surface being sprayed by means of a siphon-jet arrangement at the gun nozzle. Because the burning gases extend well beyond the nozzle, the ThermoSpray process requires that the distance to plastic base materials be increased over that shown in the figure and/or that increased surface speed be used to prevent deforming the plastic.

Plasma spraying utilizes a technique involving the application of electrical energy to ionize an inert gas (see Fig. 2.13). This creates a higher energy level which generates temperatures as high as 30,000°F (16,650°C). The primary value of such high temperatures is that the melting point of the material being sprayed is reached quickly. Thus, it need not remain in the hot zone as long as that corresponding to lower-temperature oxyacetylene flames. Requiring a shorter time at heat means it can be propelled faster through the plasma arc, to reach the substrate with less heat loss, and with greater impact. These two advantages make plasma coatings denser, and give them higher bond strengths.

The plasma gun generates its high temperatures without combustion, so even the most reactive materials can be melted with little or no oxidation, and without altering the original chemical composition to any significant degree. Also, the spraying can be done within a protective atmosphere chamber, to further protect the sprayed material. The elimination of oxides produces a more cohesive coating which will finish to a better surface and protect itself better in service.

Most metals can be sprayed on to almost any material. The substrate need not be heated over 300°F, (most materials are unaffected using a 400°F maximum), to eliminate heat distortion problems.

Three of the manufacturers of the flame-spray process of metalizing are:

Metco, Inc.
1101 Prospect Ave.
Westbury, Long Island, N.Y. 11590
Phone: (516) 334-1300

Metalizing Co. of America, Inc.
3520 West Carroll Ave.
Chicago, Ill. 60624
Phone: (312) 722-3710

Wall Colmonoy Corp.
19345 John R. Street
Detroit, Michigan 48203
Phone: (313) 893-3800
TWX: (810) 243-4048

2.1.3 SMALL-APERTURE METALS

Metals such as screens, metal mesh, conductive textiles, mesh-laminated optical display windows, and honeycomb, which are called in this handbook *small-aperture metals* (SAM) do not follow the shielding effectiveness relations of Sec. 1.4. In fact, even if the bonding of the joints of screens and mesh were perfect, the shielding effectiveness in terms of the resistance per square follows a different relationship at higher frequencies. At HF and above there is little evidence of the relationship of the base material conductivity and permeability of SAM to the shielding effectiveness in contrast to both the homogeneous and PHM. Except for some metalized textiles, the SAM family may have typically from 10 to 1000 times as much air path as basic metal, i.e., the base metal fibers or wires occupy about 0.1 to 10% of the optical path.

2.1.3.1 WIRE SCREENS

As suggested above, the SAM family is characterized by metal-thickness-to-air-gap area ratios of the order of 0.05 (may range from 0.01 for chicken wire to more than 0.1 for some screens and approximately 0.5 for some metalized textiles). The air-gap spacings, g, may vary from about 20 mils (0.51 mm) to 2 inches (5.1 cm). Above low frequency, the shielding effectiveness model which fairly describes this family in the far field (plane waves) is:

$$SE_{dB} \simeq 20 \log_{10}\left(\frac{\lambda/2}{g}\right) \text{ dB for } g \leq \lambda/2 \tag{2.8}$$

$$\simeq 0 \text{ for } g \geq \lambda/2 \tag{2.9}$$

where, λ = wavelength
g = air gap dimension in same units as λ

Thus, the SAM family may be described as incremental waveguides below cutoff and the shielding effectiveness is due to reflection loss with essentially no loss due to absorption.

Eq. (2.8) may be expressed in both the English and metric systems for frequency instead of wavelength, i.e.:

$$SE_{dB} \simeq 20 \log_{10}\left(5906/g_{in} f_{MHz}\right) \text{ dB, for } g_{in} \leq 5906/f_{MHz} \tag{2.10}$$

$$\simeq 0 \text{ for } g_{in} \geq 5906/f_{MHz} \tag{2.11}$$

$$\simeq 20 \log_{10}\left(15000/g_{cm} f_{MHz}\right) \text{ dB, for } g_{cm} \leq 15{,}000/f_{MHz} \tag{2.12}$$

$$\simeq 0 \text{ for } g_{cm} \geq 15{,}000/f_{MHz} \tag{2.13}$$

As the frequency continues to decrease, S_{dB} does not rise at 20 dB/decade indefinitely for far-field conditions. Measurements indicate that a maximum of about 110 dB (5-1/2 decades below cutoff = $\lambda/2$) results for copper and aluminum screen and upwards of 140 dB for galvanized steel. To support this theory it appears that when the screen mesh gap, g, becomes a very tiny fraction of a wavelength (i.e., 10^{-5} to 10^{-7}), the material appears to approach a homogeneous media from a macroscopic point-of-view. To the extent that this applies, the relations of Sec. 1.4 may once again govern for electric and magnetic field ($R \ll \lambda/2\pi$) and plane waves ($R \gg \lambda/2\pi$) with the further restriction that $g <$ about $10^{-6}\lambda$*.

Eqs. (2.10) through (2.13) are plotted in Fig. 2.14. The figure shows that for plane waves the shielding effectiveness of screen becomes relatively small above a few GHz whereas it is very significant below 1 MHz. The shielding effectiveness to electric fields ($R \ll \lambda/2\pi$ would be greater than shown in Fig. 2.14 whereas to magnetic fields ($R \ll \lambda/2\pi$) it would be less. It is assumed here that the crossing screen mesh wires are making good electrical contact with each other by welds or other bonding processes.

2.1.3.2 PANELS WITH HOLE APERTURES

While not a member of the wire and mesh screen family, metal panels having punched holes are often found in many applications. Convection air cooling of metal boxes containing electronic parts is one use in which the top cover plate frequently is perforated with small round apertures.

One shielding effectiveness relation for low-impedance, magnetic fields reported in the literature** for square cover plates having round perforations, is:

$$SE_{dB} = 32t/g + 4 + 20 \log_{10}\left[(D/g)^3/N\right]^{***} \tag{2.14}$$

where, terms are shown in Fig. 2.15 insert

In examining Eq. (2.14) or Fig. 2.15, it is immediately evident that it is independent of frequency. The development of Sec. 2.1.3.1 and other sections clearly indicates that a frequency-dependent relation should exist. Since this is an unknown limitation on Eq. (2.14), until a better mathematical model is developed, it is recommended that:

---

* Design curves similar to those in the appendices could be developed here, but insufficient measurement data exist to support this at the time of writing this handbook.

** See Chap. 4 Reference: AFSC Design Handbook, *DH 1-4 Electromagnetic Compatibility*.

*** See formula (11.7) and (11.8) in Handbook Volume 3.

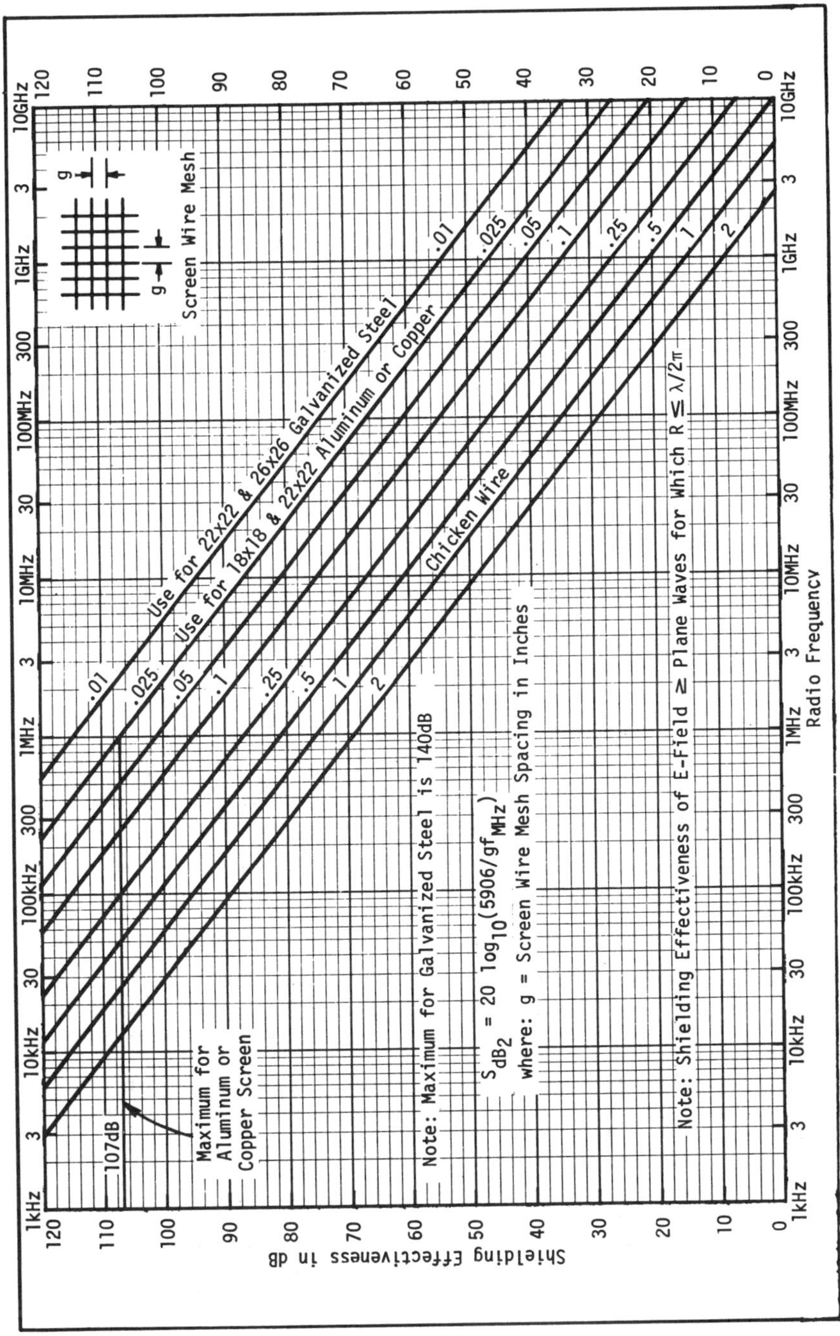

Figure 2.14 - Shielding Effectiveness of Screen Wire to Plane Waves

(1) Compute SE_{dB} by Eq. (2.14) or Fig. 2.15

(2) Compute SE_{dB} by Eq. (2.8) or Fig. 2.14

(3) Select the smaller of #1 or #2 for SE_{dB} to be used

Illustrative Example 2.1

A 1/16 inch thick (1.59 mm) cover plate is to be secured with gasketing on top of a five-sided aluminum box containing a 30 MHz-250 MHz surveillance receiver. The cover plate measures 19" x 14" and the perforation array of holes measures 16" (40.64 cm) x 12" (30.48 cm). The holes are to be punched 1/4" (6.35 mm) in diameter with a one inch (25.4 mm) center-to-center spacing. Compute the (worst) shielding effectiveness to low-impedance, magnetic fields at 250 MHz:

$$D = \sqrt{D_n \cdot D_w} = \sqrt{16 \times 12} = 13.86"$$

$$D/g = 13.86/0.25 = 55.4$$

$$t/g = (1/16)/0.25 = 0.25$$

$$N_h = D_n/1" + 1 = 17$$

$$N_w = D_w/1" + 1 = 13$$

$$N = N_h \cdot N_w = 221 \text{ holes}$$

From Eq. (2.14):

$$SE_{dB} = 32(0.25) + 4 + 20 \log_{10}(55.4)^3/221 = 69.7 \text{ dB}$$

From Fig. 2.14: SE_{dB} = 39 dB. Thus, selecting the smaller of the two, yields SE_{dB} = 39 dB. The two expressions would become equal when frequency in Fig. 2.14 ≃ 7 MHz. Below this frequency, Eq. (2.14) would govern as long as it does not exceed Fig. 2.14.

2.1.3.3 METALIZED TEXTILES

Metalized textiles represent an interesting development in recent years regarding what can be done with fabrics to convert them into electromagnetic shields for people, areas, and equipment. Among the shielding applications of metalized textiles are partial shields for plastic equipment cases, shielding and grounding curtains, body contacts for electromedical sensors, static discharge wiper-elements, latching devices, protective clothing for high-voltage linemen, flexible-shielded shrouds and boots, and EMI-shield accordian-type walkways.

One family of conductive textile products is the silver metalized, woven-nylon fabrics made by Technical Wire Products (part Nos. 77-09775 to -09778). The pure silver coating of the nylon cloth substrates provides surface resistances of less than 10 Ω/sq. High strength, light

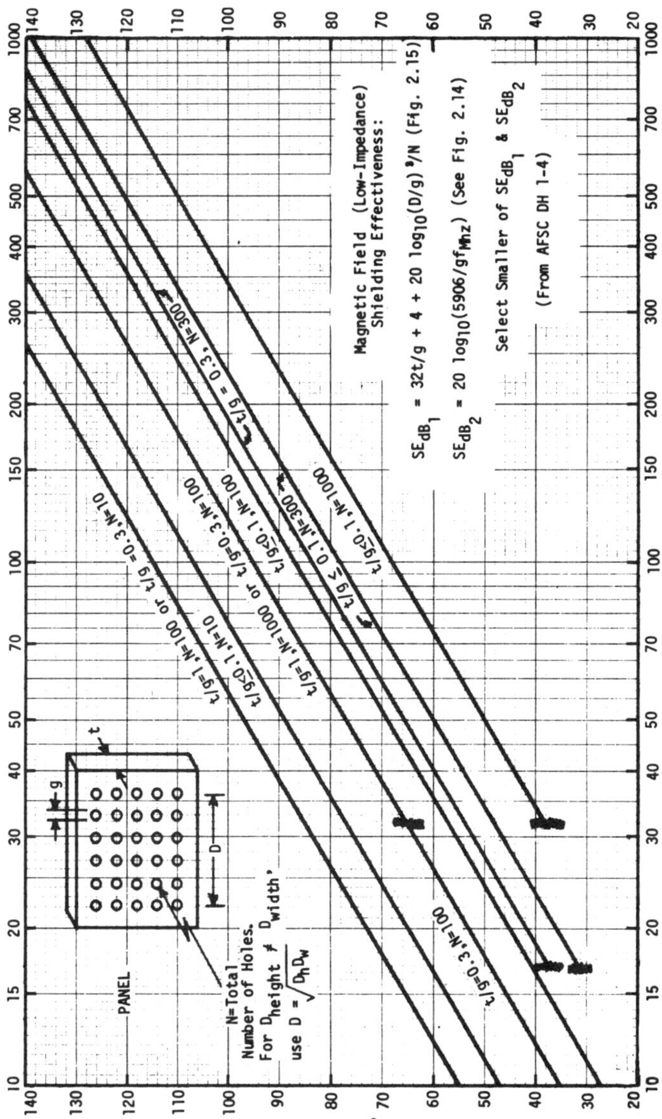

Figure 2.15 - Shielding Effectiveness, SE_{dB_1}, of Panel with Holes

weight and resistance to moisture and chemicals is inherent in the nylon fabrics employed as the substrates.

The measured shielding effectiveness of the metalized textiles to plane waves follows a curve approximated by g ≈ 0.03 in Fig. 2.14. The maximum low-frequency, plane-wave shielding, however, ranges from about 65 dB for the heavy marquisette weave (Tecknit #77-0977) to about 150 dB for the very fine rip-stop weave (Tecknit #77-09776).

Textile widths of 36" (91 cm) are available in rolls up to 200 ft. (61 m) long. Also available are 0.5 mil (13 μm) pressure sensitive, polyester fiber for easy installation and 1 mil clear nylon to insulate or protect the textile surface. To maintain performance when exposed to the elements, a vinyl-coated fabric is available to weatherproof the fabric and bind the yarn together. The rated operating temperature range is from -60°F (-51°C) to 250°F (121°C).

2.1.3.4 MESH-LAMINATED OPTICAL DISPLAY WINDOWS

An open-mesh screen laminated between optical materials is used to provide both EMI shielding and visual display. The primary advantage of these windows over that of thin vacumm deposited transparent coated conductive windows (see Sec. 2.14) is their high light transmission properties. The wire mesh lamina (e.g.; Tecknit EMC-GLAS) have 90% to 95% open area while the coated windows range from 60% to 75% light transmission.

By measurement, the shielding effectiveness of the mesh windows to plane waves follows a curve approximated by g ≈ 0.25 to 0.5 in Fig. 2.14. This is about 14 dB less than expected theoretically for the open 2 mil (51 μm) diameter Sn/Cu/Fe. Measured electric-field performance is rated from 35 dB to 90 dB over the 200 Hz to 100 MHz spectrum while magnetic-field performance is rated from 10 dB to 50 dB over the 200 Hz to 10 MHz spectrum per MIL-STD-285 (see Sec. 2.2).

2.1.4 SHIELDED OPTICAL DISPLAY WINDOWS

The topic of optical display windows having extremely-thin conductive coatings for R-F shielding, is an extension of Sec. 2.1.1 involving homogeneous metals, but in which $t/\delta \ll 1$. The subject involves the use of such windows for optical viewing scope faces, LED or Nixie displays, panel meters or simply see-through applications in which radiated R-F fields must be attenuated. Sec. 11.1.3 of Vol. 3 of the EMI/EMC Handbook Series discusses shielded optical display windows in length. The discussion here is limited to shielding effectiveness.

The light transmission of conductive glass vs deposited surface resistance is shown in Fig. 2.16. Low surface resistance, while

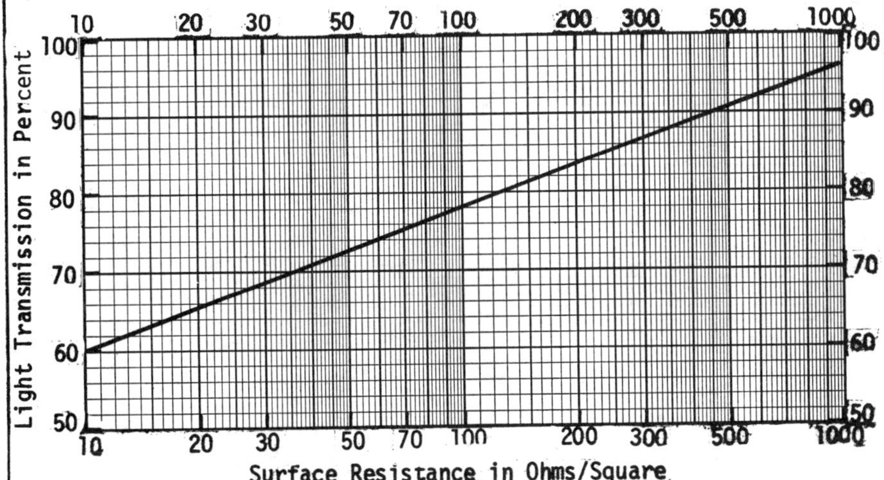

from Lassiter, H.A. re Conductive Glass - See Chap. 4, Reference

Figure 2.16 - Light Transmission of Conductive Glass

corresponding to better shielding effectiveness, is achieved at the expense of light transmission loss. Typical application trade offs correspond to a range of transmission from 60% (10 ohms/sq) to 80% (125 ohms/sq). Figs. 2.17 and 2.18 show the shielding effectiveness of various deposited thicknesses of gold vs frequency for a source-to-shield distance of 1 meter and 1 km. Gold is used because of its high stability (low on the electrochemical series) and good conductivity. Its metal cost is negligible for thicknesses below 1 µm.

Since $t \ll \delta$ in the graphs of Figs. 2.17 and 2.18, Eqs. (1.102), (1.103), (2.6), and (2.7) also apply in which:

$$SE_{dB} = 20 \log_{10}(188.5\ k/R_{sdc}) \qquad (1.103)$$

$$R_{sdc} = 17.2/\sigma_r t_{nm} = 24.6/t_{nm} \quad \text{for gold} \qquad (2.14)$$

$$= 6.79 \times 10^{-4}/\sigma_r t_{mils} = 9.7 \times 10^{-4}/t_{mils} \quad \text{for gold} \qquad (2.15)$$

Thus, $t = 1$ nm corresponds to $R_{sdc} = 24.6$ ohms/sq

$= 3$ nm corresponds to $R_{sdc} = 8.2$ ohms/sq, etc.

Figs. 2.17 and 2.18 are also labeled in units of R_{sdc} since the D-C surface resistivity provides a convenient measurement method for controling the amount of the deposition of metal consequently, a conductive display window of 80% light transmission corresponds to about $R_{sdc} = 125\ \Omega$/sq or a thickness of about 0.2 nm.

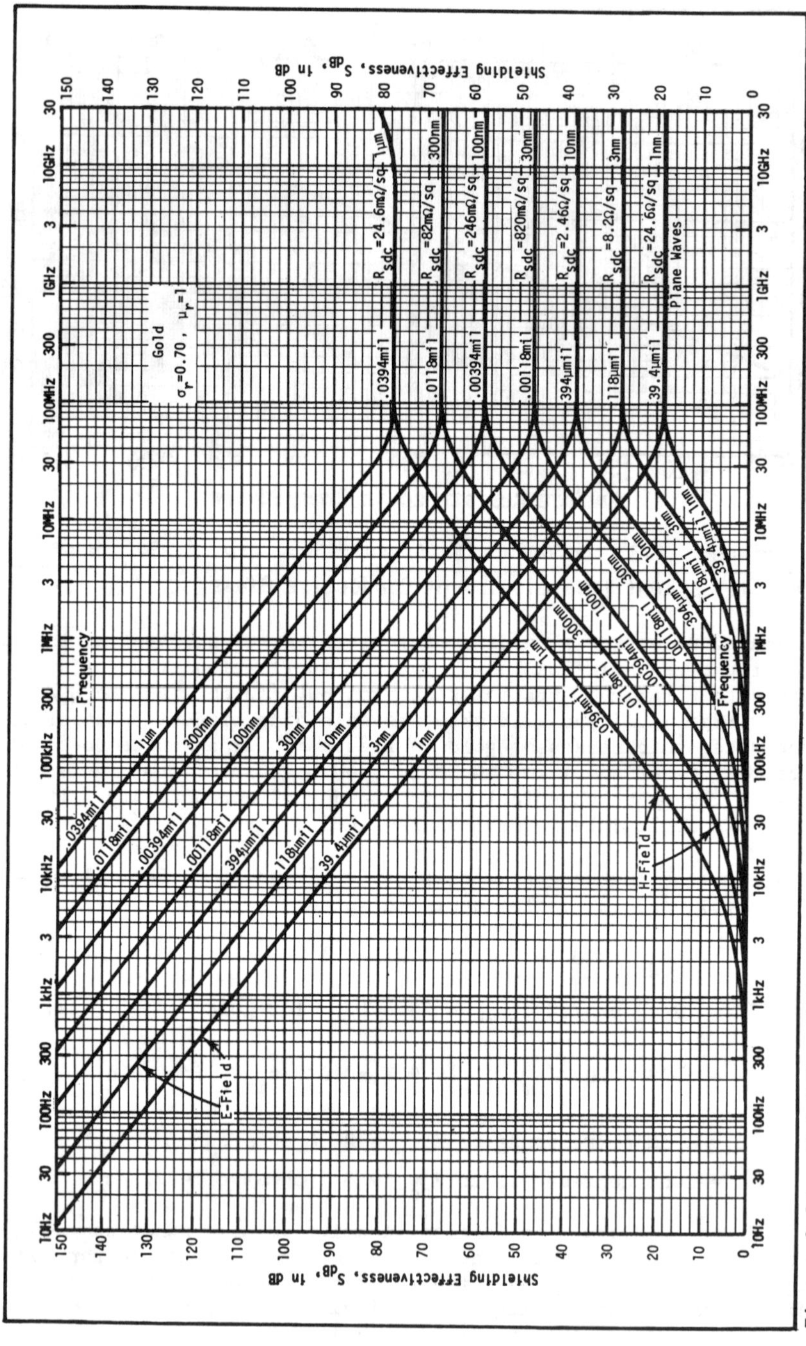

Figure 2.17 - Shielding Effectiveness of Gold vs. Frequency for Source-to-Metal Distance of 1m

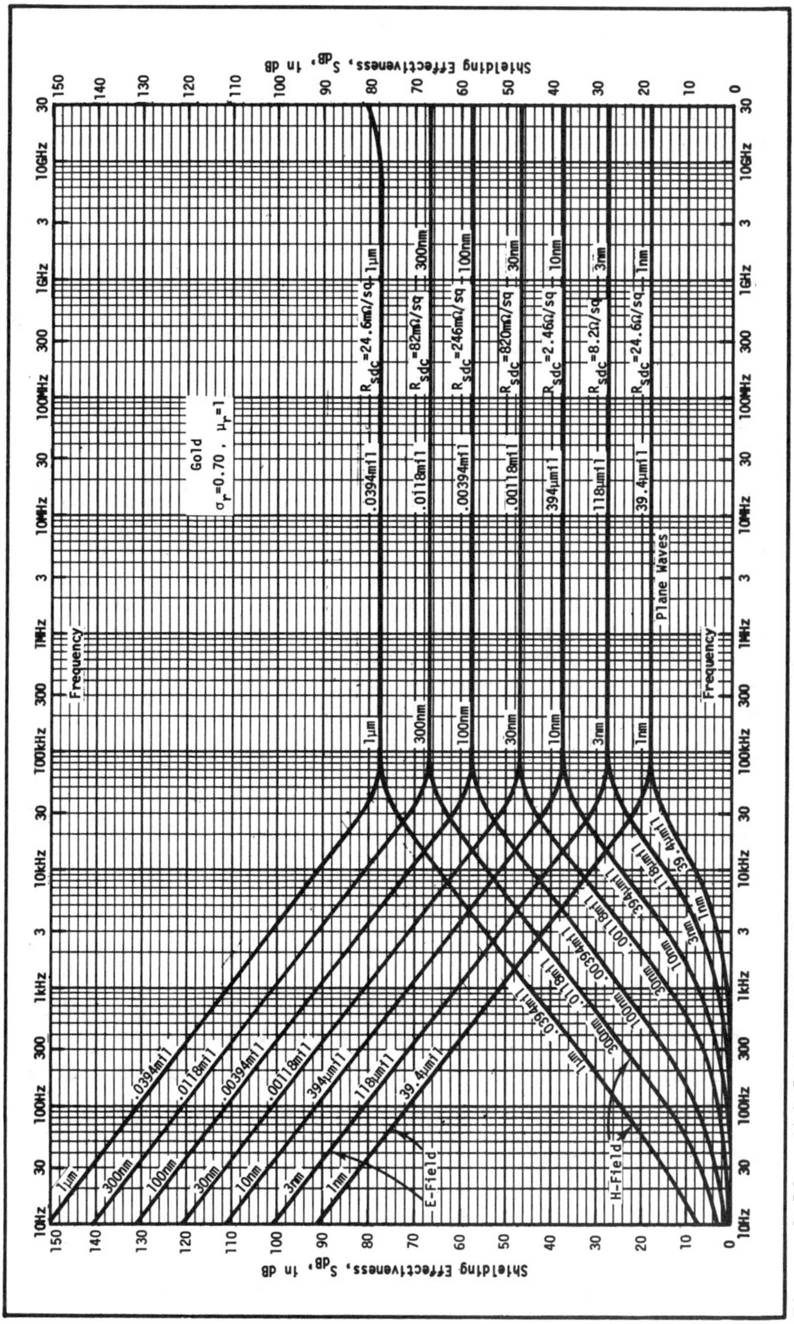

Figure 2.18 - Shielding Effectiveness of Gold vs. Frequency for Source-to-Metal Distance of 1km

2.2 SHIELDING DENSITY FOR WEIGHT-SENSITIVE APPLICATIONS

Apart from the shielding leakage problems discussed in Sec. 1.4.6, experimental data have led to the conclusion that the size or shape of a shielded enclosure does not alter its shielding effectiveness below the cavity-resonance frequencies (see Sec. 1.4.6.2).

For many applications, especially for aircraft and spacecraft, there exists a need for a *figure-of-merit* to relate the shielding effectiveness desired in terms of either the weight or volume of metal required. Consequently, the concept of a shielding density was devised by Schulz to relate the shielding effectiveness per unit weight per unit area of the shielding metal.

The weight per unit area (W/A) of some of the more popular metals, which may be considered for use in weight-sensitive applications, is listed in Tab. 2.2. When transferred to an actual enclosure size, other considerations must also be factored, as applicable, in determining the actual enclosure weight, viz: (1) the weight of entry/exit connectors must be added and the weight of punched-out metal subtracted; (2) the weight of finger stock or other gasketing, extra interlocking lips, and hinges must be added; (3) copper-plating and soldering weight must be added; (4) the reduction in weight due to bend radii over a square-cornered box must be subtracted; and (5) weight of excess metal at seams must be added due to overlaps, soldering, braizing, welding, etc. For the purpose of the present discussion, however, these weight additions and subtractions are not considered here.

Table 2.2 - Weight per Unit Area per Unit Thickness of Some Metals

Metal	$oz/ft^2/mil$	$gram/m^2/\mu m$	Relative σ_r	Relative μ_r
Copper	0.740	8.89	1	1
Monel	0.732	8.79	0.041	1
Brass	0.705	8.47	0.47	1
Steel	0.649	7.79	0.02	200
S3 Netic	0.642	7.71	0.172	300
Titanium	0.375	4.50	0.036	1
Aluminum	0.225	2.70	0.63	1
Magnesium	0.143	1.71	0.38	1

The desired figure-of-merit, designated *shielding density,* SD, is defined as follows:

$$SD = SE_{dB} \; W/A \tag{2.16}$$

where, SE_{dB} = shielding effectiveness in dB defined in Chap. 1 and the appendices

W/A = weight per unit area of metal of thickness, t,

required to yield SE_{dB} under the defined conditions.

It is noted that identical SD values for different metals *do not* have identical electromagnetic SE_{dB} values. SD values *do* indicate which metal will provide the greatest amount of shielding in relation to weight.

In considering shielding effectiveness, the worst case situation is selected, i.e., low-frequency magnetic fields with relatively little distance between emission source and areas to be protected. The rationale for this is that there will always be low-frequency magnetic-field components from a source emission unless the source is carrier operated, such as a transmitter. Further, all carrier-operated and, for that matter, any electronic equipment will have base-band amplifiers extending down to either DC or a low frequency even though this may not be the most sensitive frequency region.

In the low-frequency portion of the spectrum (approximately below 100 kHz), the shield metal thickness of non-magnetic materials is usually a small fraction of a skin depth ($t/\delta << 1$) and absorption loss can be ignored. This was developed for Eqs. (1.100) and (1.102):

$$SE_{dB} \approx 20 \log_{10}(0.707kt/\delta), \text{ for } t/\delta << 1 \text{ and } K >> 1 \qquad (1.100)$$

$$= 20 \log_{10}(188.5k\sigma t) \qquad (1.102)$$

$$= 20 \log_{10}\left(5.82\text{x}10^{-3} r_m f_{Hz} \sigma_r t_{mils}\right) \qquad (2.17)$$

$$= -44.7 + 20 \log_{10}\left(r_m f_{Hz} \sigma_r t_{mils}\right) \qquad (2.18)$$

for magnetic fields and $r_m \leq \lambda/2\pi$

where, r_m = emission-source to metal-shield distance in meters

f_{Hz} = frequency in Hertz

σ_r = conductivity of metal relative to copper

t_{mils} = metal thickness in mils

Solving Eq. (2.18) for the metal thickness, yields:

$$t_{mils} = \frac{\log_{10}^{-1}\left(\frac{SE_{dB}+44.7}{20}\right)}{r_m f_{Hz} \sigma_r} \qquad (2.19)$$

The weight associated with this metal required to give a shielding effectiveness of SE_{dB} is:

$$W/A = (W/A/mil) t_{mils} = \frac{(W/A/mil) \log_{10}^{-1}\left(\frac{SE_{dB}+44.7}{20}\right)}{r_m f_{Hz} \sigma_r} \qquad (2.20)$$

where, (W/A/mil) is obtained from Tab. 2.2. The conditions under which Eqs. (2.19) and (2.20) apply ($t/\delta \ll 1$ and $K \gg 1$) are:

(1) $r_m < \lambda/2\pi$ (in the near field) (2.21)

(2) $5000 \geq t_{mil}\sqrt{f_{Hz}\sigma_r\mu_r}$ (absorption loss is negligible) (2.22)

(3) $SE_{dB} \geq 10$ dB (2.23)

Eqs. (2.19) and (2.20) are plotted for aluminum in Fig. 2.19 vs frequency with SE_{dB} and r_m as parameters. The graph lines are coded in accordance with the listings in Tab. 2.3. This table is used to select the applicable line in the figure.

Table 2.3 - Applicable Line Selection for Use in Fig. 2.19

SE_{dB} \ r_m	1cm	1.8cm	3.2cm	5.6cm	10cm	18cm	32cm	56cm	1m	1.8m	3.2m	5.6m	10m	18m
10dB	M	L	K	J	I	H	G	F	E	D	C	B	A	
15	N	M	L	K	J	I	H	G	F	E	D	C	B	A
20	P	N	M	L	K	J	I	H	G	F	E	D	C	B
25	Q	P	N	M	L	K	J	I	H	G	F	E	D	C
30	R	Q	P	N	M	L	K	J	I	H	G	F	E	D
35		R	Q	P	N	M	L	K	J	I	H	G	F	E
40			R	Q	P	N	M	L	K	J	I	H	G	F
45				R	Q	P	N	M	L	K	J	I	H	G
50					R	Q	P	N	M	L	K	J	I	H
55						R	Q	P	N	M	L	K	J	I
60							R	Q	P	N	M	L	K	J
65		Region of No Validity						R	Q	P	N	M	L	K
70									R	Q	P	N	M	L
75	Viz, 5000 is not $\geq t_{mil}\sqrt{f_{Hz}\sigma_r\mu_r}$									R	Q	P	N	M
80											R	Q	P	N
85												R	Q	P
90													R	Q
95														R
100														

Illustrative Example 2.2

Determine the thickness and weight of sheet aluminum stock required to give 30 dB of shielding effectiveness at 400 Hz when the magnetic emission source is: (1) 10 cm away, and (2) 3 m away from the location of the planned shield.

From Tab. 2.3, use line M for r_m = 10 cm and line G for r_m = 3 m in Fig. 2.19. At 400 Hz, Fig. 2.19 indicates:

(1) 10 cm distance: $t \gg 100$ mils (2540 μm) and $W/A \gg 22.5$ oz/ft^2 (6858 grams/m^2).

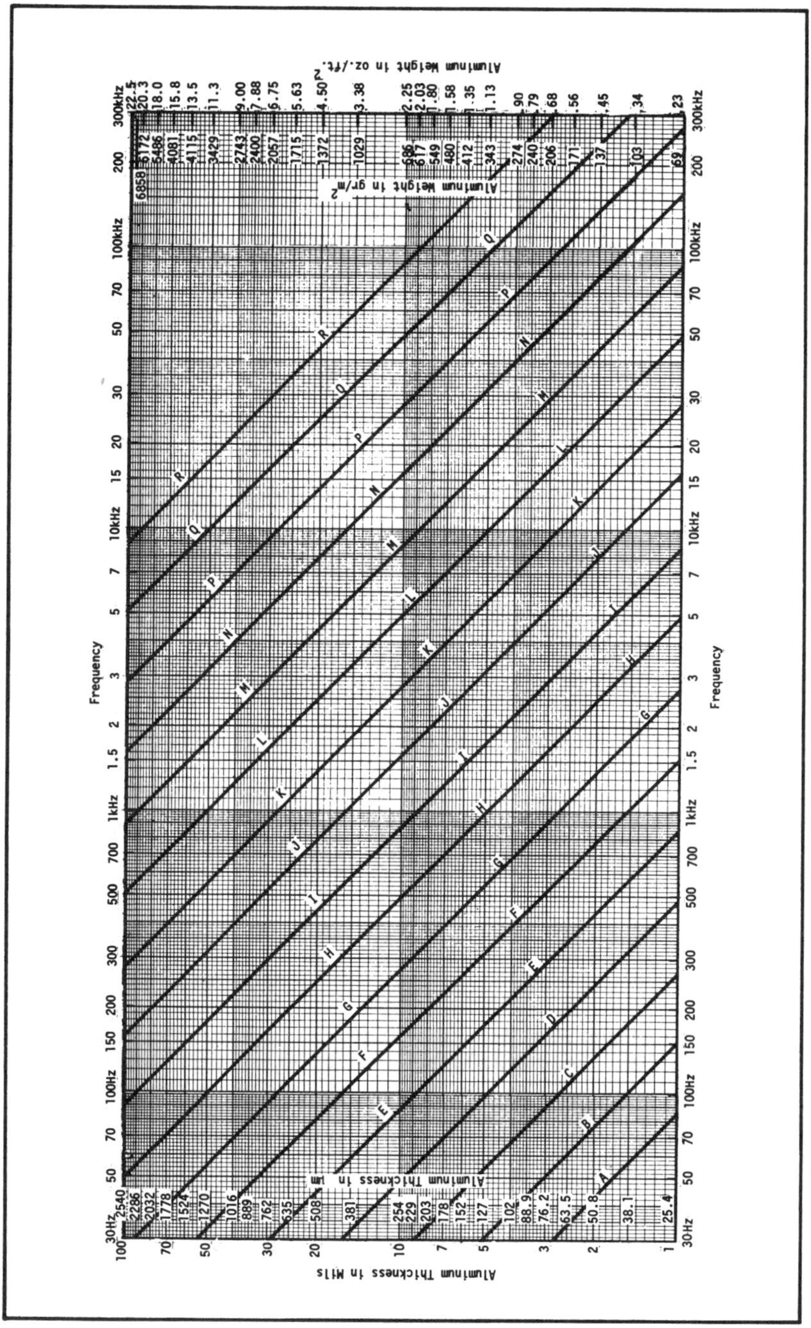

Fig. 2.19- Aluminum Thickness and Weight vs. Frequency

(2) 3 m distance; t = 6.8 mils (175 μm) and W/A ≃ 1.55 oz/ft^2 (475 grams/m^2).

Thus, aluminum will not work for magnetic sources which are only 10 cm away since the shield would be too thick and heavy for weight-sensitive applications.

It remains to compare the options in the above example for metals other than aluminum in order to determine if a lighter metal can be found with the same shielding effectiveness. The answer is obtained by using Eqs. (2.19) and (2.20) in which the ratio of t_{mil} is formed from the two (or more) candidate metals:

$$\left(t_{mil}\right)_{ratio} = \frac{\sigma_{r1}}{\sigma_{r2}} \tag{2.24}$$

$$(W/A)_{ratio} = \frac{(W/A/mil)_2\ \sigma_{r1}}{(W/A/mil)_1\ \sigma_{r2}} \tag{2.25}$$

Eqs. (2.24) and (2.25) are computed for Tab. 2.4 relative to copper by using the information in Tab. 2.2.

Table 2.4 - Relative Thicknesses and Weights of Some Metals for Yielding the Same Shielding Effectiveness (see Constraints)

Metal	$\left(t_{mil}\right)_{ratio}$	$(W/A/mil)_{ratio}$	$(W/A)_{ratio}$
Copper	1	1	1
Monel	24.4	0.989	24.1
Brass	2.13	0.953	2.03
Steel	50	0.877	43.9
S3 Netic	5.81	0.868	5.05
Titanium	27.8	0.507	14.1
Aluminum	1.59	0.304	0.483
Magnesium	2.63	0.193	0.509

The above table shows that steel is a very poor low-frequency magnetic shield as long as it is operated under $t/\delta \ll 1$ conditions in which the absorption loss is negligible. Of course, if these condition do not apply, then their shielding effectiveness will increase substantially (see Chap. 3). Monel gives poor shielding performance with regard to both relative thickness and weight requirements. The performance of titanium is not much better. While magnesium has a lighter specific weight than aluminum, a greater thickness is required due to its lesser conductivity. For the same shielding effectiveness, the net weight is greater than aluminum although only by 5.4%. Thus, since the shielding density of aluminum and magnesium are so close, the preferred metal of the two should be chosen on a basis other than its shielding properties.

Illustrative Example 2.3

Select the metal and compute the thickness and weight required to give 50 dB of shielding effectiveness at 10 kHz to a magnetic-field source at a distance of 30 cm.

From Tab. 2.3, select line P. In Fig. 2.19 this requires at 10 kHz a thickness of aluminum of 28 mils and a weight of 6.3 oz/ft^2. Next, consider mumetal which has an σ_r = .029 and μ_r = 20,000. Since it violates the absorption loss constraint (see condition Eq. (2.22), the above curves cannot be used. From Chap. 3 and the Appendices, however, the thickness is calculated to be about 3 mils. Based on a unit weight of 0.645 oz/ft^2/mil, the total weight of the mumetal is 4 oz/ft^2 or about 63% of aluminum. Thus, the illustration shows only the very high permeable metals will give a better magnetic shielding density than aluminum for the stated conditions.

Illustrative Example 2.4

Repeat example 2.3 for a required attenuation of 30 dB with the other conditions remaining the same. Here, curve K is used in Fig. 2.19 to give an aluminum thickness of 2.8 mils and a weight of 0.63 oz/ft^2. From the HP-65 card program in Chap. 3 or from the Appendices, the required mumetal thickness is 3.7 mils and the weight is 2.39 oz/ft^2 or about 280% more weight than aluminum. The reason mumetal is heavier here is that less of the mumetal's performance is based on absorption loss than in the preceding example where greater shielding effectiveness was required (50 dB vs 30 dB). Also, note the weight saving between the two examples: reducing the required SE_{dB} from 50 dB to 30 dB results in a weight reduction to about one-sixth of the original weight. This clearly shows the importance of *not overdesigning for EMC especially for critical weight situations such as spacecraft.*

The preceding two examples clearly show the need for separate calculations for situations in which a single variable change results in opposite decisions of metal choice. Generalizations cannot be made in which both non-magnetic and magnetic metals may be used. Of the non-magnetic family, aluminum or magnesium result in the least weight for the required shielding under all conditions.

2.3 MIL-STD-285

MIL-STD-285, dated 25 June 1956, is entitled, *Military Standard Attenuation Measurements for Enclosures, Electromagnetic Shielding, for Electronic Test Purposes, Method of.* This standard is reproduced as Appendix H for the convenience of those readers interested in the details. There are several points that are mentioned here regarding MIL-STD-285 and its use.

- E-field or high-impedance waves, where $r < \lambda/2\pi$ and $Z_W >> 377$ ohms, are measured at a distance of 12 inches (30.5 cm) only from the shielding-barrier interface. This results in much higher shielding effectiveness (optimistic values) than would exist at greater distances. Measurements are made at least at 200 kHz, 1 MHz and 18 MHz and generally at other frequencies below 30 MHz.

- H-field or low-impedance waves, where $r < \lambda/2\pi$ and $Z_W << 377$ ohms, are measured at a distance of 12 inches (30.5 cm) only from the shielding-barrier interface. This results in much lower shielding effectiveness (pessimistic values) than would exist at greater distances. Measurements are made at least at 200 kHz and generally at other frequencies below 1 MHz.

- Plane Waves or far-field measurements, where $r > \lambda/2\pi$ and $Z_W = 377$ ohms, are made at a distance of six feet (1.83 m) or more (where $r \geq 2\lambda$) from the shielding barrier interface. Here distance is unimportant as long as $r > \lambda/2\pi$ except for the measurement attenuation range available. Measurements are made at least at 400 MHz and generally at several other frequencies above 100 MHz.

- Measurements are usually made out in the open first with no barrier between transmitting-receiving antennas; then they are repeated with the same orientation and separation of antennas, but with the shielding barrier located in between. This gives erroneous readings for E-fields since the metal barrier detunes the antennas by capacitive loading. Antenna probes, whose dimensions are small compared to 12 inches, will not have this problem. H-field antennas are Faraday shielded.

- The measurement range suggested in MIL-STD-285 can be substantially increased by using power amplifiers or power oscillators on the transmitting side and narrowband tuned amplifiers and/or correlators with fiber optic reference path on the receiving side. Manufacturers' E-field shielding effectiveness data often reflect inadequate measurement range.

- For gaskets, wire screens, woven mesh, and other metal thickness which is very small compared to the shielded enclosure metal thickness, an 18 inch (45.7 cm) or 24 inch (61 cm) square aperture is usuall cut in the enclosure wall of an internally partitioned shielded enclosure. Shielding effectiveness (insertion loss) measurements are made by taking the difference between field strength measurements when no test sample is used vs when the test sample is in place.

Discussions regarding test methods and procedures of MIL-STD-285 are beyond the scope of this handbook. Some considerations regarding accuracy of measurement are suggested in Chap. 5 of Vol. 2 of the EMI/EMC Handbook Series. However, it is informative to the readers here to amplify the comments in the first and second paragraphs above regarding optimistic and pessimistic shielding effectiveness data reported in manufacturers' literature. For example, suppose there is a known source of interference in a real problem at a distance of 10 meters against which one desires to shield. How much shielding effectiveness is needed when manufacturers' data sheets are reported in the language of MIL-STD-285 at a distance of 12 inches (30.5 cm)?

The answer to the above question is to develop correction data based on extrapolating the results of the MIL-STD-285 data distance to the desired user distance. Such extrapolation data are a function of many variables and would require many graphs. However, the problem may be delimited by certain assumptions so that quantitative correction data can be presented here for user application.

For conditions under which the reflection loss is significant (i.e., $K \geq 10$), Eq. (1.64) becomes:

$$\simeq 20 \log_{10} (0.707Kt/\delta) \text{ for } t/\delta \ll 1 \qquad (1.100)$$

$$SE_{dB} \simeq 20 \log_{10} \left(e^{t/\delta} K/4\right) \text{ for } t/\delta \gg 1 \qquad (1.101)$$

where, $K = Z_w/Z_m = k \times \text{constant}$ (see Eq. (1.74))

$k = \lambda/2\pi r$ for E-fields

$= 2\pi r/\lambda$ for H-fields

$= 1$ for plane waves

When Eq. (1.100) or (1.101) is applied for MIL-STD-285 for any two distances an error is developed which is a function of k alone. Thus, for any measurement distance, r_m, and any user applied distance of r_u, the correction in shielding efficiency, ΔSE_{dB}, becomes:

$$\Delta SE_{dB} = 20 \log_{10}\left(r_m/r_u\right) \quad \text{for E-fields} \qquad (2.26)$$

$$= 20 \log_{10}\left(r_u/r_m\right) \quad \text{for H-fields} \qquad (2.27)$$

$$= 0 \quad \text{for plane waves} \qquad (2.28)$$

in which it is understood that *both* r_m and r_u are in the near field.

When one of the distances is in the near field and the other is in the far field, Eqs: (2.26) and (2.27) become:

$$\Delta SE_{dB} = 20\ \log_{10}\left(2\pi r_m/\lambda\right) \text{ for E-fields, and } r_m \text{ in near and } r_u \text{ in far field} \quad (2.29)$$

$$= 20\ \log_{10}\left(\lambda/2\pi r_u\right) \text{ for E-fields, and } r_u \text{ in near and } r_m \text{ in far field} \quad (2.30)$$

$$= 20\ \log_{10}\left(2\pi r_u/\lambda\right) \text{ for H-fields, and } r_u \text{ in near and } r_m \text{ in far field} \quad (2.31)$$

$$= 20\ \log_{10}\left(\lambda/2\pi r_m\right) \text{ for H-fields, and } r_m \text{ in near and } r_u \text{ in far fields} \quad (2.32)$$

Eqs. (2.26) through (2.32) are plotted in Fig. 2.20 for MIL-STD-285 distance of r_m = 0.305 m(12 inches) and any user distance, r_u, as shown

Illustrative Example 2.5

A manufacturer's literature states that per MIL-STD-285 a metalize silicone elastomer offers at least 90 dB of shielding effectiveness to E-fields at 10 MHz, 30 dB shielding effectiveness to H-fields at 10 MHz and 70 dB to plane waves above VHF. Determine the likely shielding effectiveness at a distance of 10 meters.

From Fig. 2.20, the correction to E fields at r_u = 10 m is -23 dB and to H fields is + 23 dB. Thus, for a 10 m distance SE_{dB} (E-field) = 67 dB and SE_{dB} (H-field) = 53 dB. Note that the 10 m distance is in the far field ($r_u > \lambda/2\pi$) at 10 MHz, since the r_u = 10 m line in Fig. 2.2 ends at 4.8 MHz ($r_u = \lambda/2\pi$). Thus, there is no correction for SE_{dB} (plane waves) = 70 dB.

Illustrative Example 2.6

The shielding effectiveness to H-fields of two mils (51 μm) sheet foil of aluminum was measured at a 1 m distance at 42 kHz and found to be 50 dB (see curve M, Fig. 2.19). What protection would be offered by this foil at a 5 cm distance from a hostile magnetic field?

Since both conditions are in the near field, Eq. (2.27) applies, $\Delta SE_{dB} = 20\ \log_{10}(5cm/100cm) = -26$ dB. Thus, SE_{dB} (H-field) = 50 dB - 26 dB = 24 dB.

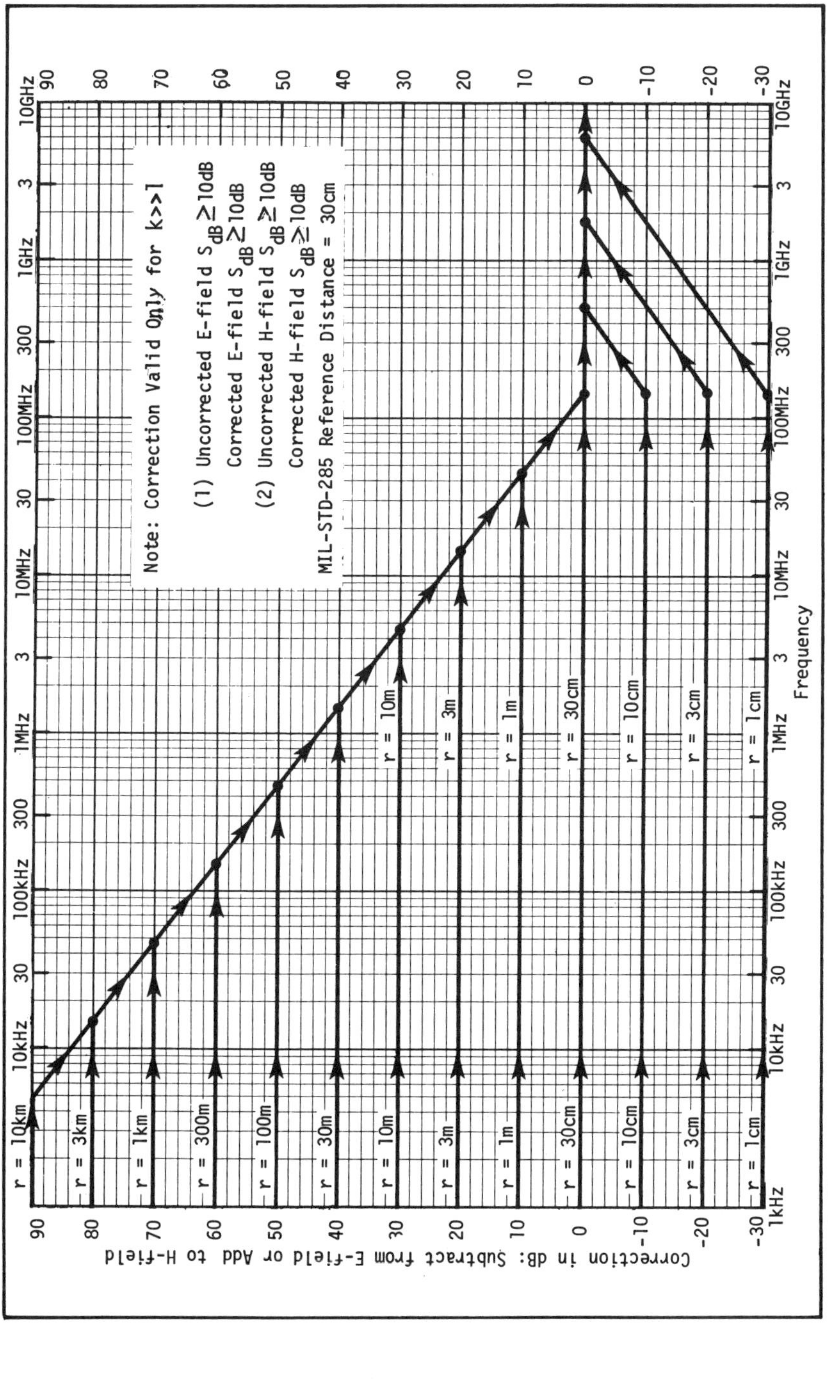

Figure 2.20- Correction in Shielding Effectiveness to Convert MIL-STD-285 Results to Another Distance

CHAPTER 3

APPLICATIONS AND EXAMPLES

This chapter discusses how to use the shielding design graphs in this handbook with particular emphasis on the graphs presented in the appendices. A number of illustrative examples is given to familiarize the reader with the design graphs and to exercise the variables involved. Finally, the last section in this chapter presents a Hewlett-Packard HP-65 programmable calculator program for those who wish to compute the shielding effectiveness of any metal for any applicable input data.

3.1 HOW TO USE THE DESIGN GRAPHS

Chaps. 1 and 2 contain 25 graphs of data pertinent to the topic of electromagnetic shielding. However, for most shielding design situations, the 42 graphs presented in the appendices are of more direct use in computing shielding effectiveness vs:

(1) frequency
(2) conductivity of metal relative to copper
(3) permeability of metal relative to copper
(4) metal thickness
(5) distance between emission source and metal shield
(6) E-field, H-field, or plane-wave emission source

Each of the 42 graphs contains an enormous amount of data. The dependent variable (Y-axis) is shielding effectiveness (ranges from 0 to 300 dB) and the independent variable (X-axis) is frequency which ranges from 10 Hz to 30 GHz. The variable parameters are metal thickness (ranges from 0.0001 mil or 2.54 nm to 1 inch or 2.54 cm) and source-emission field type (E-field/plane wave or H-field/plane wave). Each graph corresponds to a metal class (7 different combinations of relative conductivity and permeability) and to a source-to-metal distance range (10 cm to 10 km). Thus, 7 x 6 or 42 graphs in this handbook are required to define the shielding effectiveness of any metal.

It becomes necessary to classify each metal in order to determine which graph set to use. Thus, that metal closest to the conductivity and permeability of one of the seven sets is captioned by the set type. While this may result in an error in shielding effectiveness of a few to several dB in some instances, it otherwise prevents the requirement for hundreds of graphs corresponding to all specific metals. The metal class set (actually the appendix letter nomenclature) is defined in Tab. 3.1. Tab. 3.2 uses the metals listed in Tab. 2.1 and the definition in Tab. 3.1 to establish which appendix to use for any listed meta

Table 3.1 - Definition of Metal Class Based on Permeability and Conductivity Relative to Copper

Class or Metal Type (Appendix letter)	Lower $\sqrt{\mu_r\sigma_r}$	Upper $\sqrt{\mu_r\sigma_r}$	Lower $\sqrt{\mu_r/\sigma_r}$	Upper $\sqrt{\mu_r/\sigma_r}$
A = Copper	1	1	1	1
Class A	0.45	>1	1	2.24
B = Monel	0.2	0.2	5.0	5.0
Class B	0.05	0.45	2.24	20
C = Nickel	4.7	4.7	21.3	21.3
Class C	<4.7	8	<21.3	70
D = Iron	13.0	13.0	76.7	76.7
Class D	<4.7	19	55	325
E = Hypernick	12.5	12.5	357	357
Class E	7	19	325	650
F = 78 Permalloy	29.4	29.4	272	272
Class F	19	40	140	650
G = High Permeability	56.6	56.6	1414	1414
Class G	24	>56.6	650	>1414

As explained in Sec. 2.1.1, the permeability of a magnetic metal is frequency and flux-density dependent. Thus, it would seem to be impossible to define shielding effectiveness without knowing operating conditions. Consequently, the applicable appendix in Tab. 3.2 shows a frequency split at 300 kHz (see Fig. 2.1) above which $\mu_r \to 1$ for magnetic materials. Accordingly, above 300 kHz for the higher conductivity ($\sigma_r \geq 0.2$) magnetic metals, Appendix A (copper) is used, while for $\sigma_r < 0.2$ magnetic metals, Appendix B (monel) is used.

After selecting the appendix letter to use from either Tabs 3.1 or 3.2, as applicable, it remains to select the correct design graph within the appendix. Each appendix subset is based on the distance between the EMI emission source and the metal barrier. Six such subsets are presented in each appendix as shown in Tab. 3.3 (the X corresponds to any appendix letter, viz., A.B.C, etc.).

Having selected the applicable graph within the appendix, it remains to apply the graph to the problem. The graph may be used to determine either: (1) the resulting shielding effectiveness given the metal thickness, operating frequency, and E or H-field problem, (2) the required metal thickness given the desired shielding effectiveness,

Table 3.2 - Metal Class for Choice of Appendix to Use

Metal	Appendix No. < 300kHz	Appendix No. ≥300kHz	Metal	Appendix No. <300kHz	Appendix No. ≥300kHz
Alfenol, 16	E	B	Mumetal	G	B
Aluminum, soft	A		Nichrome	B	
Aluminum, tempered	A		Nickel	C	A
Beryllium	A		Permalloy,45	D	A
Brass, (91% Cu, 34% Z_n)	A		Permalloy,78	F	B
Brass, (66% Cu, 34% Z_n)	A		Permalloy,4-79	G	B
Cadmium	A		Permendur	D	A
Chromium	A		Perminvar,45-25	C	B
Conetic	G	B	Platinum	B	
Copper	A		Silver	A	
Gold	A		Sinimax	E	B
Hypernick	E	B	Steel,cold-rolled	C	B
Hiperco	D	B	Steel,Hot-rolled	D	B
Iron, commercial	C	B	Steel,50% Nickel	D	B
Iron, purified	F	B	Steel,stainless	D	B
Iron, Silicone 4%	D	B	Supermalloy	G	B
Lead	B		Tantalum	B	
Magnesium	A		Tin	B	
Manganese	B		Tungsten	A	
Mercury	B		Zinc	A	
Monel	B				
Monimax	D	B			

Table 3.3 - Applicable Appendix Design Graph for Specified Distance

Figure Number	Nominal Distance	Applicable For
X.1	R_m = 10cm	3cm ≤ R_m ≤ 30cm
X.2	R_m = 1m	30cm ≤ R_m ≤ 3m
X.3	R_m = 10m	3m ≤ R_m ≤ 30m
X.4	R_m = 100m	30m ≤ R_m ≤ 300m
X.5	R_m = 1km	300m ≤ R_m ≤ 3km
X.6	R_m = 10km	R_m ≤ 3km

operating frequency, and E or H-field problem, or (3) the upper and lower useful frequency given the desired shielding effectiveness, the metal thickness, and E or H-field problem.

3.2 ILLUSTRATIVE EXAMPLES

This section presents a number of examples which serve to illustrate how the design graphs in the appendices can be used.

Example #1

A sheet of 1/16 inch (i.e., 62.5 mils or 1.59 mm) iron is being considered for shielding a box from a strong magnetic field originating from a 60-Hz generator located 5 feet (1.59 m) away. Determine the shielding effectiveness in dB.

Unless the type of iron is specified (see Tab. 3.2), Appendix D is selected as listed in Tab. 3.1. Since the distance is 1.52 m, Tab. 3.3 indicates that Fig. D.2, within Appendix D is the closest to operational conditions. Select the design curve of t = 100 mils (to start) at 60 Hz for H-field conditions. Interpolating between t = 30 mils and t = 100 mils reveals that the expected shielding effectiveness is about 24 dB. Were the design objective to provide a shielding effectiveness of 50 dB at this distance, Fig. D.2 indicates that the thickness of the iron should be about 140 mils (3.6 mm).

Example #2

A piece of sensitive electronic equipment is located near (100 m) an A-M broadcast station antenna transmitting at 1250 kHz. Determine the shielding effectiveness of a 1/32 inch (i.e., 31.25 mils or 0.79 mm) sheet metal aluminum box enclosure to E-fields or plane waves, as applicable.

Tab. 3.2 indicates that Appendix A applies. For a distance of R_m = 100 m in Tab. 3.3, the closest graph is identified as A.4. Corresponding to a metal thickness of 30 mils and a frequency of 1.25 MHz, the shielding effectiveness for plane waves in the figure is about 196 dB. Note that 196 dB would never be obtained in practice because of the penetrations required into and out of the box (see Chap. 11, Vol. 3, EMC Handbook Series). Also note that the near/far-field interface ($R_m = \lambda/2\pi$) exists at 38.2 m. Thus, the electronic equipment box housing is located in the far field of the transmitter and plane-wave conditions apply rather than E-field conditions.

Were the design objective to provide a shielding effectiveness of 100 dB at the 100-m distance, Fig. A.4 indicates that the thickness of aluminum should be about 0.7 mil (18 μm). Thus, a 1 mil foil covering the box would be adequate notwithstanding the fact that it would offer no mechanical strength. Note that graphs A.5 (1 km) and A.6 (10 km) also yield the same results since all are in the far field at 1250 kHz.

Example #3

A ground-level nuclear detonation produces a broadband electromagnetic pulse (EMP) with most of its energy distributed in the 10 kHz band. From a distance of 5 km, the blast center looks like a magnetic source having billions of amperes. To protect electromagnetic equipment, specify the thickness of both aluminum and sheet steel to provide 300 dB isolation.

Tabs. 3.2 and 3.3 indicate that graphs A.6 (aluminum) and D.6 (steel) apply. For frequency = 10 kHz, the thickness of aluminum should be about 700 mils (18 mm) and that of steel should be about 50 mils (1.3 mm). A separate calculation of magnetic-flux density should be made to determine if the steel is saturated (i.e., the permeability is lowered).

Example #4

A nearby (500 m) L-Band weather radar operating at 1300 MHz creates an electric-field strength of 100 V/m in an internal room in a building where a computer is to be located. It is known that the computer, its peripherals, and all interconnecting cables will not be susceptible to radiation below 3 V/m. Thus, determine a relatively inexpensive shield for the room.

The required shielding effectiveness is the ratio of the electric-field strengths which equals about 31 dB. Consider the shielding effectiveness of 1 mil (25.4 μm) household aluminum foil as one possible solution. Here the walls, ceiling, and flooring would be so lined with overlapping joints taped in place. Tabs. 3.2 and 3.3 show that graph A.5 of Appendix A applies. For a frequency of 1300 MHz, the shielding effectiveness is about 200 dB. Thus, the aluminum foil provides an enormous shielding effectiveness and it remains to insure that door seams, power entrance, and the like do not create the mode of EMI entry.

Example #5

A cable is carrying a number of 10 kb/sec digital data lines at 5V logic and 100-ohm impedance. The differential-mode current is 5V/100Ω = 50mA. At a distance of 5cm, the magnetic-field intensity from a single wire is 0.16 A/m and the magnetic-flux density is $B = \mu H = 2$ mgauss = 2×10^5pT = 106 dBpT. If the differential-mode reduction due to reverse field gradients and twisted-wire pairs is about 70 dB, calculate the required thickness of a 10 cm pipe of permeable material to reduce the flux density to 1 milligamma (0 dBpT) on the outside.

The required shielding effectiveness of the pipe is 106 dBpT - 70 dBpT = 36 dB. This minimum attenuation must be provided from DC to about 10 kHz (upper frequency for 50 μsec pulses; $f \approx 1/\pi t$). Since the least pipe attenuation exists at the lower-frequency components, a frequency of 10 Hz will be used which is the lower limit on the appendix graphs.

For iron (class D) and a distance of R_m = 5 cm, Tabs. 3.2 and 3.3 indicate that graph C.1 in the appendices should be used. The graph shows that a wall thickness of about 150 mils (4 mm) is required. Had mumetal (graph G.1) been used, the required wall thickness would be reduced to about 10 mils (.25 mm). Note that the shielding effectiveness of these metals is lowest at 10 Hz (about 36 dB) whereas at 10 kHz it is >300 dB for the iron and 88 dB for the mumetal.

3.3 HP-65 PROGRAM FOR SHIELDING EFFECTIVENESS

Most of the graphs presented in this handbook were developed by programming pertinent mathematical models on the Hewlett-Packard Model HP-65 programmable calculator. This calculator uses a magnetic card to transfer up to 100 steps of stored program. The HP-65 contains four working registers (the stack) and nine storage registers. It was selected since it has adequate capacity to run off many of the shielding relations and it is readily available in USA at a cost of about $700 as of 1975.

Specifically, the shielding effectiveness relation presented in Eq. (1.64) is the main program. The supporting relations used in the program are given in Eqs. (1.68), (1.71), (1.74) and (1.77). It took quite a bit of program manipulation to get all the relations into a 100-step program to meet the limitations of the HP-65. This program is presented in Table 3.4. Thus, those having access to the HP-65 need merely to key in the 100-step program of Tab. 3.4 and transfer memory to a permanent magnetic card.

3.3.1 USER PROGRAM INSTRUCTIONS

The *User Program Instructions* are listed below. From the magnetic card developed from Tab. 3.4, load the HP-65 with this Shielding Effectiveness Program #1513. Next, key in the pertinent input data and store in registers as follows:

Key In	Explanation	Register Store
f_{MHz}	frequency in MHz	#1
σ_r	conductivity relative to copper	#2
μ_r	permeability relative to air	#3
t_{min}	metal thickness in mils (1 mil=25.4 micrometer)	#4
R_m	source-to-shield distance in meters	#5
1 or 2	#1 = E-fields or plane waves #2 = H-fields or plane waves	#6

To see the shielding effectiveness, key user label "A". The HP-65 takes about 10 seconds to compute the answer in dB.

If it is desired to see a new shielding effectiveness based on new input data, it is *not* necessary to key in registers #1 through 6 all over again. It is only necessary to input one or more registers based on one or more new data and then key label "A" again.

Table 3.4 - HP-65 Shielding Effectiveness Program Steps

Card # 1513 Program. Title: SHIELDING EFFECTIVENESS

KEYS	CODE
LBL	23
A	11
g	35
RAD	42
4	04
8	08
RCL 1	3401
÷	81
RCL 5	3405
÷	81
STO 7	3307
1	01
gx>y	3524
1	01
E	15
RCL 6	3406
1	01
gx=y	3523
RCL 7	3407
E	15
RCL 7	3407
g	35
1/x	04
LBL	23
E	15
EEX	43
6	06
X	71
RCL 2	3402
RCL 1	3401
÷	81
RCL 3	3403
÷	81
f	31
$\sqrt{x}$	09
X	71
STO 8	3308
RCL 4	3404
.	83
7	07

KEYS	CODE
7	07
X	71
RCL 1	3401
RCL 2	3402
X	71
RCL 3	3403
X	71
f	31
$\sqrt{x}$	09
X	71
CHS	42
STO 7	3307
RCL 8	3408
1	01
−	51
f^{-1}	32
$\sqrt{x}$	09
RCL 7	3407
f^{-1}	32
LN	07
X	71
CHS	42
RCL 7	3407
gx≶y	3507
f^{-1}	32
R→ρ	01
RCL 8	3408
1	01
+	61
f^{-1}	32
$\sqrt{x}$	09
+	61
f	31
R→ρ	01
4	04
÷	81
RCL 8	3408
÷	81
RCL 7	3407
2	02

KEYS	CODE
÷	81
CHS	42
f^{-1}	32
LN	07
X	71
g	35
ABS	06
f	31
LOG	08
2	02
0	00
X	71
R/S	84
LBL	23
B	12
3	03
9	09
4	04
X	71
STO 4	3304

Notes:

Registers:

R1 Frequency in MHz	R4 t in mils	R7 working
R2 σ_r	R5 R_m in meters	R8 working
R3 μ_r	R6 E or H, 1 or 2	R9

3.3.2 ILLUSTRATIVE EXAMPLES

The following examples will serve to illustrate the use of the HP-65 shielding effectiveness program. The results may be compared with those in the appendices* or in Sec. 3.2.

Example #1

A sheet of 1/16 inch (i.e., 62.5 mils) iron is being considered to shield a box from a strong magnetic field originating from a 60-Hz generator located 5 feet (1.52 meters) away. Compute the shielding effectiveness in dB.

After loading the magnetic program card into the HP-65 calculator, key in f_{MHz} = 60 x 10^{-6} MHz = 60 EEX 6 CHS (STO 1), σ_r = .17 for iron (STO 2), μ_r = 1000 for iron (STO 3), t = 62.5 mils (STO 4), R_m = 1.52 meters (STO 5), and 2 for H-field (STO 6). To see shielding effectiveness, key label "A" and see 24 dB.

Example #2

A piece of sensitive electronic equipment is located near (100 m) an A-M broadcast station antenna transmitting at 1250 kHz. Determine the shielding effectiveness of a 1/32 inch (31.25 mils) sheet metal aluminum box enclosure to E-fields or plane waves, as applicable.

After loading the magnetic program card, key in f_{MHz} = 1.25 MHz (STO 1), σ_r = .61 for aluminum (STO 2), μ_r = 1 (STO 3), t = 31.25 (STO 4), R_m = 100m (STO 5), and 1 for E-field or plane wave (STO 6). To see shielding effectiveness, key label "A" and see 196 dB. Note that 196 dB would never be obtained in practice because of the penetrations required into and out of the box (see Chap. 11, Vol. 3 EMC Handbook Series). Also note that the near/far-field interface ($R_m = \lambda/2\pi$) exists at 38.2 m. Thus, the box is located in the far field of the transmitter and plane-wave conditions apply rather than E-field conditions.

Example #3

A ground-level nuclear detonation produces a broadband electromagnetic pulse (EMP) with most of its energy distributed in the 10 kHz band. From a distance of 5 km, the blast center looks like a magnetic source having billions of amperes. To protect electromagnetic equipment, specify the thickness of both aluminum and sheet steel to provide 300 dB isolation.

Key f_{MHz} = .01 MHz (STO 1), σ_r = .61 (STO 2), μ_r = 1 (STO 3), t

* If you like this program, contact Don White Consultants, Inc. for other HP-65 EMC programs.

(this is a guess) = 250 mils (STO 4), R_m = 5000 m (STO 5), and H/PW = 2 (STO 6). Key "A" to see SE_{dB} = 191 dB. Since this is inadequate, try t = 1 inch (1000 mils - STO 4), and again key "A" to see 387 dB. Again try t = 800 mils (STO 4) and key "A" to see 334 dB. For t = 700 mils, it is seen that SE_{dB} = 308 dB.

For the sheet steel, key σ_r = .17 (STO 2), μ_r = 1000 (STO 3) and t (this is a guess) = 100 mils (STO 4). Key "A" to see SE_{dB} = 526 dB. Next try t = 1/16 inch (62.5 mils) to see SE_{dB} = 362 dB. For t = 50 mils, the shielding effectiveness is 308 dB, the same as 700 mils of aluminum.

Example #4

A nearby (500 m) L-Band weather radar operating at 1300 MHz creates an electric-field strength of 100 V/m in an internal room in a building where a computer is to be located. It is known that the computer, its peripherals, and all interconnecting cables will not be susceptible to radiation below 3 V/m. Thus, determine a relatively inexpensive shield for the computer room.

The required shielding effectiveness is the ratio of the electric-field strengths or 100/3 which equals about 31 dB. Consider the shielding effectiveness of 1 mil (25.4 μm) household aluminum foil as one possible solution. Here the walls, ceiling, and flooring would be so lined with overlapping joints taped in place. Key f_{MHz} = 1300 MHz (STO 1), σ_r = .61 (STO 2), μ_r = 1 (STO 3), t = 1 mil (STO 4), R_m = 500 m (STO 5), and E/PW = 1 (STO 6). Key "A" to see SE_{dB} = 169 dB. Thus, the aluminum provides an enormous shielding effectiveness and it remains to insure that door seams, power entrance, and the like do not create the mode of EMI entry.

CHAPTER 4

REFERENCES

(1) Abramowitz, M.; Stegun, I.A., "Handbook of Mathematical Functions", Boulder, Colorado; National Bureau of Standards; 1964.

(2) Adams, W.S., "Graphical Presentation of Electromagnetic Shielding Theory", *Proceedings Tenth Tri-Service Conference on Electromagnetic Compatibility;* Chicago, Ill., Armour Research Foundation; pp. 421-499; November 1964.

(3) AF/BSD Exhibit 62-87, "Electro-Interference Control Requirements for Minuteman (WS 133B)", 6 December 1962.

(4) Albin, A.L., "Optimum Shielding of Equipment Enclosures", *Electronic Design,* Vol. 8, No. 3; February 3, 1960.

(5) "An Investigation into Existing Calibration Methods of Field Intensity Meters", Rome Air Development Center, Rome, New York; Tech. Report RADC-TR-64-527, January 1965.

(6) Bannister, P.R., "The Quasi-near Fields of Dipole Antennas", *IEEE Transactions on Antennas and Propagation;* Vol. AP-15; pp 618-626; September 1967.

(7) Bannister, P.R., "New Theoretical Expressions for Predicting Shielding Effectiveness for the Plane Shield Case", *IEEE Transactions on Electromagnetic Compatibility;* Vol. EMC-10, No. 1; March 1968.

(8) Blewett, J.P., "Magnetic Field Configuration Due to Air Core Coils", *John Hopkins Applied Physics;* Vol. 18, p. 968; November 1947.

(9) Boeing Aircraft Corporation D2-2-2444, "Electro-Interference Control Requirements (Equipment)", 5 March 1959.

(10) Bozorth, R.M., "Ferromagnetism", Princeton, New Jersey; *Von Nostrand,* 1951.

(11) Bridges, J.E., Huenemann, R.G., and Hegner, H.R., "Electric Field Shielding and Measurement", *IEEE Electromagnetic Compatibility*

Symposium Record, Vol. 27C80; Washington, D.C., July 18-20, 1967; pp. 173-177.

(12) Bridges, J.E., "Proposed Recommended Practices for the Measurement of Shielding Effectiveness of High-Performance Shielding Enclosures", *IEEE Transactions on Electromagnetic Compatibility;* Vol. EMC-10, No. 1; P. 82; March 1968.

(13) Bridges, J.E., Miller, D.A., and Schulz, R.B., "Comparison of Shielding Calculations", *IEEE EMC Transactions;* Vol. EMC-10, No. 1; March 1968; p. 175.

(14) Bridges, J.E., and Zalewski, R.A., "Magnetic Field Pickup by Flexible Braid Coaxial Cables", *IEEE Transactions on Electromagnetic Compatibility;* Vol. EMC-10, No. 1; March 1968; p. 130.

(15) Bronwell, A.B., and Beam, R.E., "Theory and Application of Microwaves", McGraw-Hill, New York; 1945; pp. 337-339.

(16) Brush, D.R., Schulz, R.B., and Jorgensen, L., "Initial Magnetic Permeability Measurements of RF Shielding Materials", *Proceedings of 8th IEEE Symposium Electromagnetic Compatibility,* July 1966.

(17) Brush, D.R., Schulz, R.B., and Jorgensen, L., "Low-Frequency Electrical Characteristics of RF Shielding Materials", *IEEE Transactions on Electromagnetic Compatibility";* Vol. 10, No. 1, March 1968; pp. 67-72.

(18) Cacace, R., and Hassett, R., "Investigation of Measurement Techniques for Transient Magnetic Fields", *Symposium Digest, 7th National Symposium on Electromagnetic Compatibility;* June 28-30, 1965; New York, New York.

(19) Clough, L., and Salzetti, J., "Magnetic Induction Susceptibility at Power Frequencies", *Eighth Tri-Service Conference on EMC;* October 30 through November 1, 1962; Chicago, Ill., pp. 241-269.

(20) Cohen, D., "A Shielded Facility for Low-Level Magnetic Measurements", *Journal of Applied Physics,* Vol. 38, No. 3, 1967; pp. 1295-2196.

(21) Cole, N.H., "A Comparison of RF Shielding Materials", *Electronic Design;* Vol. 10, No. 20; Sept. 27, 1962.

(22) Collin, R.E., "Field Theory of Guided Waves", McGraw-Hill, New York, New York , 1960.

(23) "Conical Logarithmic Sprial Antennas", General Dynamics/Astronautics, San Diego, California, U.S. Air Force AF33-(616)-7436; 1963.

(24) Cooley, W.W., "Low-Frequency Shielding Effectiveness of Nonuniform Enclosures", *IEEE Transactions on Electromagnetic Compatibility;* Vol. 10, No. 1; March 1968; pp. 34-43.

(25) Cowdell, R.B., "New Dimensions in Shielding", *IEEE Transactions on Electromagnetic Compatibility;* Vol. 10, No. 1; March 1968; pp. 158-167.

(26) Cowdell, R.B., "Simplified Shielding", *1967 IEEE Electromagnetic Compatibility Symposium Record,* Vol. 27C80; Washington, D.C.; July 18-20, 1967.

(27) DeMilt, M.D., Loh, Y.P., and Showers, R.M., "Resonance Properties of the Shield of a Coaxial Cable over a Ground Plane" (Abstract); *IEEE Transactions on Electromagnetic Compatibility";* Vol. EMC-10, No. 1; March 1968; p. 135.

(28) Dinger, H.E., and Raudenbush, J., "A Technique for Measuring the Effectiveness of Various Shielding Materials", Naval Research Laboratory, NRL Report No. 4103, AD-9353, January 1953.

(29) Dolle, W.C., Van Steenberg, G.N., and Jouffray, O.L., "Effects of Shielded Enclosure Resonances on Measurement Accuracy", *Record of 1970 IEEE International Symposium on Electromagnetic Compatibility,* Vol. 70C28-EMC, July 14-16, Anaheim, 1970; pp. 417-420.

(30) Donohue, R.K., and Eckersley, A., "Some Electromagnetic Compatibility Grounding and Shielding Considerations in Titanium Aircraft Structures" (Abstract); *IEEE Transactions on Electromagnetic Compatibility;* Vol. EMC-10, No. 1, March 1968; p. 142.

(31) Dwight, H.B., "Tables of Integrals", MacMillan, 1947, N.Y.

(32) Eckersley, A., "H-Field Shielding Effectiveness of Flame-Sprayed on Thin Solid Aluminum and Copper Sheets", *IEEE Transactions on Electromagnetic Compatibility;* Vol. 10, No. 1; March 1968; pp. 101-104.

(33) "Electronics Installation and Maintenance Book-Radio Interference Reduction", Department of the Navy, Bureau of Ships; September 1963; pp. 10-3.

(34) Epstein, M., and Schulz, R.B., "Magnetic-field Pickup for Low-Frequency Radio Interference Measuring sets", *IRE Transactions, Electron Devices;* Vol. ED-8, January 1961; pp. 70-77.

(35) Erdelyi, A. Ed., "Tables of Integral Transforms (Vol. 2)", McGraw-Hill, New York; 1954.

(36) Ervin, H.W., "Shield Termination Prediction Method", *Proceedings Eighth Tri-Service Conference on Electromagnetic Compatibility,*

Armour Research Foundation, Chicago; October 1962.

(37) Farhat, N.H., Loh, Y.P., and Showers, R.M., "Effects of Partial Shields on Transmission Lines at Low Frequencies", *IEEE Transactions on Electromagnetic Compatibility;* Vol. EMC-10, No. 1; March 1968; pp. 42-51.

(38) Feber, R.R., and Young, F.J., "The Shielding of Electromagnetic Pulses by the Use of Magnetic Materials", *Record of the 1969 IEEE Symposium on Electromagnetic Compatibility;* Vol. 69C3-EMC; June 17-19, 1969, Asbury Park; pp. 73-74.

(39) Ficchi, R.F., "Electrical Interference", Hayden, New York, 1964.

(40) Filtron Company, Inc., "Interference Reduction Guide for Design Engineers", Prepared for U.S. Army Electronics Labs., Fort Monmouth, New Jersey; Vols. 1 and 2, Accession AD 619666 and AD 619667.

(41) Foster, J., Buegal, K., and Sowa, C., "Electromagnetic Compatibility Lunar Orbiter", *NASA,* Document No. NAS 1-3800; January 29, 1965.

(42) Free, W.R., "Radiated EMI Measurements in Shielded Enclosure", *Record of the 1967 IEEE Symposium on Electromagnetic Compatibility;* Vol. 27C80; Washington, D.C., July 18-20, 1967; pp. 43-53

(43) Free, W.R., et al., "Electromagnetic Interference Measurement Methods-Shielded Enclosure", Final Report on Contract No. DA 28-043 AMC-02381 (E), EES, Georgia Institute of Technology; 1967.

(44) Free, W.R., et al., "Compact Chamber for Impedance and Power Testing of VHF Whip Antennas", Final Report on Contract DAAB07-67-C-0575, EES, Georgia Institute of Technology, 1968.

(45) GM07-59-2617A, "Electro-Interference Control Requirements for Minuteman (WS 133A)", 20 October 1959.

(46) Goubau, G., "Waves on Interfaces", *Transactions IRE Antennas and Propagation,* Vol. AP-7; December 1959; pp. S140-S147.

(47) Haber, F., "Generation of Standard Fields in Shielded Enclosures", *Proceedings of IRE,* Vol. 42, November 1954.

(48) Haber, F., "Study of GSFC Radio Frequency Interference (RFI) Design Guideline for Aerospace Communication Systems", Moore School of EE, University of Pennsylvania, Report No. 66027 for NASA, April 30, 1966.

(49) Harrington, J.G., and Schulz, R.B., "Design of Minimum Weight and Maximum Effectiveness of Very-Low-Frequency Shielding", *IEEE Transactions on Electromagnetic Compatibility;* Vol. EMC-10, No. 1;

March 1968; pp. 152-157.

(50) Harris, G.R., "Precision Methods used in Constructing Electric Wave Filters for Carrier Systems", *Bell Systems Technical Journal;* Vol. 11; April 1932; pp. 264-283.

(51) Harrison, C.W., Jr., "Transient Electromagnetic Field Propagation Through Infinite Sheets into Spherical Shells and into Hollow Cylinders", *IEEE Transactions Antennas and Propagation;* Vol. AP-12; May 1964; pp. 319-334.

(52) Harrison, C.W., Jr., and Papas, C.H., "On the Attenuation of Transient Fields by Imperfectly Conducting Spherical Shells", *IEEE Transactions on Antennas and Propagation;* Vol. AP-13; November 1965; pp. 960-966.

(53) Hays, J.B., "Protecting Communications Systems from EMP Effects of Nuclear Explosions", *IEEE Spectrum;* Vol. 1; May 1964; pp. 115-122.

(54) Hollway, D.L., "Screen Rooms and Enclosures", *Proceedings IREA,* Vol. 21, No. 10; October 1960; pp. 660-668.

(55) Hornbostel, D.H., and Becker, S., et. al., "Study of Thin-film Interference Reduction Techniques", AIL Division Cutler-Hammer, Deer Park, New York; USAECOM Contract DA28-043-AMC-01970(E), 3rd Quarterly Report; July 1967. (Also, 1st and 2nd Quarterly Reports. Final Report, May 1966 to October 1967).

(56) Jackson, J.D., "Classical Electrodynamics", Wiley, New York; 1962.

(57) Jahnke, E., and Emde, F., "Tables of Functions with Formulas and Curves", Dover, New York; 1945.

(58) Jarva, W., "Shielding Efficiency Calculation Methods for Screening Waveguide Ventilation Panels and other Perforated Electromagnetic Shields", *Proceeding 7th Tri-Service Conference on Radio Interference Reduction and Electronic Compatibility;* Armour Research Foundation; November 1961; pp. 478-498.

(59) Johnson, W.R., et. al., "Development of a Space Vehicle Electromagnetic Interference Compatibility Specification", *NSA,* Contract No. 9-73-5, *TRW,* 08900-60001-T000; June 28, 1968.

(60) Jordan, E.C., "Electromagnetic Waves and Radiating Systems", Prentice-Hall, Englewood Cliffs, New Jersey; 1950.

(61) Kaden, H., "Wirbelstrome und Schirmung in der Nachrichtentechnik", Springer; Berlin, 1959; pp. 71-85, 93-107.

(62) Kanellakos, D.P., and Schulz, R.B., "New Techniques for Evaluating the Performance of Shielded Enclosures", *Fifth Conference on Radio Interference Reduction and Electronic Compatibility;* Armour Research Foundation, IIT, Chicago, Ill.; October 1959.

(63) Karakash, J.J., "Transmission Lines and Filter Networks", MacMillan, New York; 1950.

(64) King, L.V., "Electromagnetic Shielding at Radio Frequencies", *Phil. Mag. J. Sci.;* Vol. 15; February 1933; pp. 201-223.

(65) Klouda, J.C., "Practical Aspects in Evaluating Shielded Rooms", *Electro-Technology;* June 1961.

(66) Kozakoff, D.J., Bolt, A.T., and Howard, F.D., "Shielding Effectiveness of Various Shaped Geometrical Enclosures in Terms of Normalized Parameters", *Record of the 1970 IEEE Regional Electromagnetic Compatibility Symposium Record;* Vol. 70C64-REGEMC; San Antonio, October 6-8, 1970; pp. V-A-1 to V-A-11.

(67) Larmor, J., "Electromagnetic Induction in Conducting Sheets and Solid Bodies", *London, Edinburgh, and Dublin Phil. Mag. and J. Science;* Ser. 5, Vol. 17; January 1884.

(68) Lassiter, H.A., "Low Frequency Shielding Effectiveness of Conductive Glass", *IEEE Transactions Electromagnetic Compatibility;* Vol. EMC-6; July 1964; pp. 17-30.

(69) Levy, S., "Electromagnetic Shielding Effect of an Infinite Plane Conducting Sheet between Coaxial Cables", *Proceedings of IRE;* Vol. 21; June 1963; pp. 923-941.

(70) Liaw, Y.S., "Calculation of Shielding Effectiveness of Cylindrical Cable at Low Frequencies", M.S. Thesis, Moore School of Electrical Engineering, University of Pennsylvania, Philadelphia; 1967.

(71) Lien, R.H., "Radiation from a Horizontal Dipole in a Semi-infinite infinite Dissipative Medium", *John Hopkins Applied Physics;* Vol. 24; January 1953; pp. 1-4.

(72) Lindgren, E.A., "Contemporary RF Enclosures", Erik A. Lindgren and Associates; 1967.

(73) Lockwood, R.O., "New Technique for the Determination of the Integrity of Shielded Enclosures by the Measurement of the Perpendicular Magnetic Field", *Record of the 1967 IEEE Symposium on Electromagnetic Compatibility;* Vol. 27C80; Washington, D.C., July 18-20; pp. 61-69.

(74) Loh, Y.P., "Shielding Theory of Coaxial Cylindrical Structures", Vol. EMC-10, No. 1; March 1968; pp. 16-28.

(75) Lyons, W., "Experiments on Electromagnetic Shielding at Frequencies Between One and Thrity Kilocycles", *Proceedings of IRE;* Vol. 21, April 1933; pp. 574-590.

(76) Maxwell, J.C., "On the Induction of Electric Currents in an Infinite Plane Sheet of Uniform Conductivity", *London, Edinburgh, and Doublin Phil. Mag. and J. Science (Proc. Royal Soc.);* Ser. 4, Vol. 43, No. 289.

(77) McAdam, W., and Vandeventer, D., "Solving Pickup Problems in Electronic Instrumentation", *ISA Journal;* Vol. 7, No. 4; Apr. 60; p. 48.

(78) McDonald, G.M., and Taylor, G.R., "Shielding Grounding and Circuit-Grounding Effectiveness in Interference Reduction in the 50 Hz to 15 kHz Frequency Region", *IEEE Transactions on Electromagnetic Compatibility;* Vol. EMC-8, No. 1; March 1966; pp. 8-16.

(79) Mei, K., and Van Bladel, "Low-Frequency Scattering by Rectangular Cylinder", *IEEE Transactions Antennas and Propagation;* Vol. AP-11; January 1963; pp. 52-56.

(80) Meindl, J.D., "The Calculation of Field External to Shielded Transmission Lines", Ph. D. dissertation, Carnegie Institute of Technology; Cleveland, Ohio, 1958.

(81) Mendez, H.A., "A New Approach to Electromagnetic Field-Strength Measurements in Shielded Enclosures", *Convention Record of the 1968 WESCON,* Technical Papers Session 19, Los Angeles, Calif.; August 20-23, 1968.

(82) Mendez, H.A., "Meaningful EMC Measurements in Shielded Enclosures", *Record of the 1969 IEEE Symposium on Electromagnetic Compatibility;* Vol. 69C3-EMC, Asbury Park; June 17-19, 1969; p. 137.

(83) Metcalfe, R.E., Von Allmen, R.H., and Caprio, S.T., "Investigation of Spectrum Signature Instrumentation", *IEEE Transactions on Electromagnetic Compatibility;* Vol. EMC-7, June 1965; pp. 218-232.

(84) Miedzinski, J., "Electromagnetic Screening Theory and Practice", British Electrical and Allied Research Assn., Technical Report M/T 135; 1959.

(85) Miller, D.A., and Bridges, J.E., "Geometrical Effects on Shielding Effectiveness at Low Frequencies", *IEEE Transactions on Electromagnetic Compatibility;* Vol. EMC-8, December 1966; pp. 174-187.

(86) Miller, D.A., and Bridges, J.E., "Review of Circuit Approach to

Calculate Shielding Effectiveness", *IEEE Transactions on Electromagneti Compatibility;* Vol. 10, No. 1; March 1968; pp. 52-62.

(87) MIL-STD-285, "Method of Attenuation Measurements for Enclosures Electromagnetic Shielding for Electronic Test Purposes".

(88) Moorthy, S.C., "Coupling Between Coaxial Cables at VLF (0-50Kc/s Moore School of Electrical Engineering, University of Pennsylvania, Philadelphia; Contract NObsr 85170, Index S-FO 121511, 1965.

(89) Morecroft, J.H., and Turner, A., "The Shielding of Electric and Magnetic Fields", *Proceedings IRE;* Vol. 13, August 1925; pp. 477-505.

(90) Moser, J.R., "Low-Frequency Shielding of a Circular Loop Electro magnetic Field Source", *IEEE Transactions Electromagnetic Compatibility* Vol. EMC-9; March 1967; pp. 6-18.

(91) Moser, J.R., "An Empirical Study of ELF and VLF Shield Cans", *IEEE Transactions on Electromagnetic Compatibility",* Vol. 10, No. 1; March 1968; pp. 112-125.

(92) Osburn, J.D., Morris, F.J., "Problems Encountered During the Design and Fabrication of an ELF-VLF-Shielding Enclosure", *Record of the International IEEE Symposium on Electromagnetic Compatibility;* Vol. 70C28-EMC, Anaheim, July 14-16, 1970; pp. 472-478.

(93) O'Young, S.L., Goldman, R., and Jorgensen, L., "Survey of Techniques for Measuring RF Shielding Enclosures", *IEEE Transactions on Electromagnetic Compatibility;* Vol. 10 No. 1; March 1968; pp. 72-81.

(94) Patton, B.J., "Magnetic Shielding", Socony Mobil Oil Co., Dallas, Texas.

(95) Patton, B.J., "Room-Size Enclosure for Geomagnetic Shielding", *Record of the 1970 IEEE International Symposium on Electromagnetic Compatibility;* Vol. 70C28-EMC; Anaheim, July 14-16, 1970; pp. 89-96.

(96) Patton, B.J., and Fitch, J.L., "Design of a Room-size Magnetic Shield", Socony Mobil Oil Co., Dallas, Texas.

(97) Pearlston, C.B., "Enclosure Shielding in Radio Interference", *IEEE Third National Symposium on RFI;* Washington, D.C., June 1961.

(98) Pearlston, C.B., "Case and Cable Shielding, Bonding and Grounding Considerations in Electromagnetic Interference", *IRE Transactions on Radio Frequency Interference;* Vol. RFI-4, No. 3; October 1962; pp. 1-6.

(99) Plantz, V.C., Schulz, R.B., and Goldman, R., "Shielding Density:

An Electromagnetic Shielding Design Concept for Weight Sensitive Applications", *IEEE Transactions on Electromagnetic Compatibility;* Vol. 10, No. 1; March 1968; pp. 126-129.

(100) Plantz, V.C., Schulz, R.B., and Goldman, R., "Temperature Stress and Nuclear Radiation Effects on Electromagnetic Shielding" (Abstract), *IEEE Transactions on Electromagnetic Compatibility;* Vol. 10, No. 1; March 1968; p. 129.

(101) "Proceedings of the Magnetics Workshop", NASA Technical Memo 33-216; 1965.

(102) Proposed Specification for Electromagnetic Shielding of Enclosures and Buildings", Genistron, Inc.; Compton, Calif., Final Report U.S. Naval Civil Engineering Laboratory; Contract NBY-33220; September 10, 1962.

(103) Quine, J.P., "Theoretical Formulas for Calculating Shielding Effectiveness of Perforated Sheets and Wire Mesh Screens", *Proceedings of the Third Conference on Radio Interference Reduction;* Armour Research Foundation; February 1957; pp. 315-329.

(104) Ramo, S., Whinnery, J.R., "Fields and Waves in Modern Radio", 2nd Edition, Wiley, New York; 1953; p. 244.

(105) Ramo, S., Whinnery, J.R., and Van Duzer, "Fields and Waves in Communication Electronics", Wiley, New York; 1965; p. 294.

(106) Roseberry, B.E., and Schulz, R.B., "A Parallel-Strip Line for Testing RF Susceptibility", *IEEE Transactions Electromagnetic Compatibility;* Vol. EMC-7, June 1965; pp. 142-150.

(107) Russell, A., "The Effective Resistance and Inductance of a Concentric Main and Methods of Computing the ber and bei and Allied Functions", *Phil. Mag.,* Vol. 17, 1909; pp. 524-552.

(108) Ryan, C.M., "Computer Expression for Predicting Shielding Effectiveness for the Low-Frequency Plane Shield Case", *IEEE Transactions on Electromagnetic Compatibility;* Vol. EMC-10; September 1967; pp. 83-94.

(109) Salati, O.M., "Recent Developments in Interference", *IRE Transactions on Radio Frequency Interference;* Vol. RFI-4, No. 2; May 1962; pp. 24-33.

(110) Sanford, R.L., "Basic Magnetic Quantities and the Measurement of the Magnetic Properties of Materials", NBS Monograph 47, 1962.

(111) Schelkunoff, S.A., "Transmission Theory of Plane Electromagnetic Waves", *Proceedings of IRE;* Vol. 25, November 1937; pp. 1457-1492.

(112) Schelkunoff, S.A., "Electromagnetic Waves", Van Nostrand, Princeton, New Jersey; 1943; pp. 223-225, 303-306.

(113) Schor, F.W., "Measurement of RF Leakage in Multipin Electrical Connectors", *IEEE Transactions on Electromagnetic Compatibility;* Vol. 10, No. 1; March 1968; pp. 135-141.

(114) Schreiber, O.P., "Designing and Applying RFI Shields and Gaskets", *Proceedings of Second Conference on Radio Interference Reduction;* Armour Research Foundation, Chicago, Ill.; March 1956; pp. 343-359.

(115) Schreiber, O.P., "Some Useful Analogies for RF Shielding and Gasketing", *IEEE Third National Symposium on Radio Frequency Interference;* Washington, D.C., June 1961.

(116) Schulz, R.B., "Feasibility Study of Shielding Techniques", Presentation to Interference Reduction Branch, Engineering Support Services Department; U.S. Army Electronics Command, Fort Monmouth, New Jersey; May 1964; pp. 4-5.

(117) Schulz, R.B., "ELF and VLF Shielding Effectiveness of High-Permeability Materials", *IEEE Transactions on Electromagnetic Compatibility;* Vol. 10, No. 1; March 1968; pp. 95-100.

(118) Schulz, R.B., et. al., "Shielding Theory and Practice", *Proceedings, Ninth Tri-Service Conference on Electromagnetic Compatibility;* Armour Research Foundation, Chicago, Ill.; October 1963; pp. 597-636.

(119) Schulz, R.B., and Clapsaddle, R.L., "Electrocompatibility Aspects of Microelectronics", *IEEE Transactions on Electromagnetic Compatibility;* Vol. EMC-6; January 1964; pp. 37-46.

(120) Schulz, R.B., Huang, G.C., and Williams, W.L., "RF Shielding Design", *IEEE Transactions on Electromagnetic Compatibility;* Vol. 10, No. 1; March 1968; pp. 168-175.

(121) Schulz, R.B., Plantz, V.C., and Brush, D.R., "Low-Frequency Shielding Resonance", *IEEE Transactions on Electromagnetic Compatibility;* Vol. 10, No. 1; March 1968; pp. 7-15.

(122) Schwartz, R.F., "Bibliography on Radio-Frequency Shielding", Moore School of Electrical Engineering, University of Pennsylvania, Philadelphia; 1954.

(123) Shenfeld, S., "Shielding of Cylindrical Tubes", *IEEE Transactions on Electromagnetic Compatibility;* Vol. 10, No. 1; March 1968; pp.29-34.

(124) Siegel, N.S., "Near Field Coupling on Aerospace Vehicles", *1970*

IEEE Electromagnetic Compatibility Record; Anaheim, California; July 14-16, 1970; pp. 211-216.

(125) Sommerfeld, A., "On the Propagation of Waves in Wireless Telegraphy", (in German), *Ann. der Phys.;* Vol. 81; December 1926; pp. 1135-1153.

(126) Stirrat, W.A., "Antenna Model for RFI Work", *Proceedings, Ninth Tri-Service Conference on Electromagnetic Compatibility;* Chicago, Ill., 1963.

(127) Stirrat, W.A., "USA ECOM Contribution to Shielding Theory", *IEEE Transactions on Electromagnetic Compatibility;* Vol. EMC-10, No. 1, March 1968; pp. 63-66.

(128) Stratton, J.A., "Electromagnetic Theory", McGraw-Hill, New York; 1941.

(129) Stuckey, C.W., Free, W.R., and Robertson, D.W., "Preliminary Interpretation of Near-Field Effects on Measurement Accuracy in Shielded Enclosures", *Record of the 1969 IEEE Symposium on Electromagnetic Compatibility;* Vol. 69C3-EMC.

(130) Sunde, E.G., "Switching Center Shielding Against Atmospheric Induction", Bell Telephone Labs. (Unpublished Memorandum).

(131) Thomas, A.K., "Magnetic Shielded Enclosure Design in the DC and VLF Region", *IEEE Transactions on Electromagnetic Compatibility;* Vol. 10, No. 1; March 1968, pp. 142-152.

(132) Toler, J.C., and Evans, R., "Shielded Enclosure Specification in Perspective", *1970 Regional Electromagnetic Compatibility Symposium Record;* Vol. 70C64 - REGEMC; San Antonio, October 6-8, 1970; pp. III-A01-III-A-3.

(133) U.S. Naval Civil Engineering Laboratory, "Proposed Specification for Electromagnetic Shielding of Enclosures and Buildings", Final Project Report, Contract NBy-32220, Port Hueneme, Calif., 31 July 1963; pp. 97-113.

(134) Vaska, C.S., "Problems in Shielding Electronic Equipment", *Proceedings Conference on Radio Interference Reduction;* Chicago, Ill., Armour Research Foundation, December 1954; p.p. 86-103

(135) Vaska, C.S., "Theory, Design and Engineering Evaluation of Radio Frequency Shielded Rooms", Johnsville, Pa., Report NADC-EL-54129, Aug. 1956.

(136) Weber, E., "Electromagnetic Fields", Wiley, New York; 1950.

(137) Weck, R.A., "Advanced Shielding Techniques", U.S. Army Electronics Command, Fort Monmouth, New Jersey, Technical Report ECOM-2775; July 1966.

(138) Weck, R.A., "Thin-Film Shielding for Microcircuit Applications and a Useful Laboratory Tool for Plane-Wave Shielding Evaluations", *IEEE Transactions on Electromagnetic Compatibility;* Vol. 10, No. 1; March 1968; pp. 105-112.

(139) Weck, R.A., and Lump, C.J., "Thin-film Shielding Measurements", U.S. Army Electronics Command, Fort Monmouth, New Jersey, Technical Report ECOM-2798; January 1967.

(140) Weinstock, G.L., "Electromagnetic Interference Control Within Aerospace Ground Equipment for the McDonnell Phantom II Aircraft", *IEEE Transactions on Electromagnetic Compatibility;* Vol. EMC-7, No. 2, June 1965; pp. 85-92.

(141) Wheeler, H.A., "The Spherical Coil as an Inductor, Shield, or Antenna", *Proceedings IRE;* Vol. 46, September 1958; pp. 1595-1602.

(142) Whiteside, H., and King, R.W.P., "The Loop Antenna as a Probe", *IEEE Transactions Antennas and Propagation;* Vol. AP-12, May 1964; pp. 291-297.

(143) Wills, A.P., "On the Magnetic Shielding of Tri-lamellar Spherical and Cylindrical Shells", *Phys. Rev.;* Vol. 9; October 1899.

(144) Wright-Patterson Air Force Base, Ohio, "Handbook of Instructions for Aerospace Systems Design", *Vol. 4, Electromagnetic Compatibility AFSCM 80-9, Code SEG (SEPSM);* Basic Issue, 20 April, 1964.

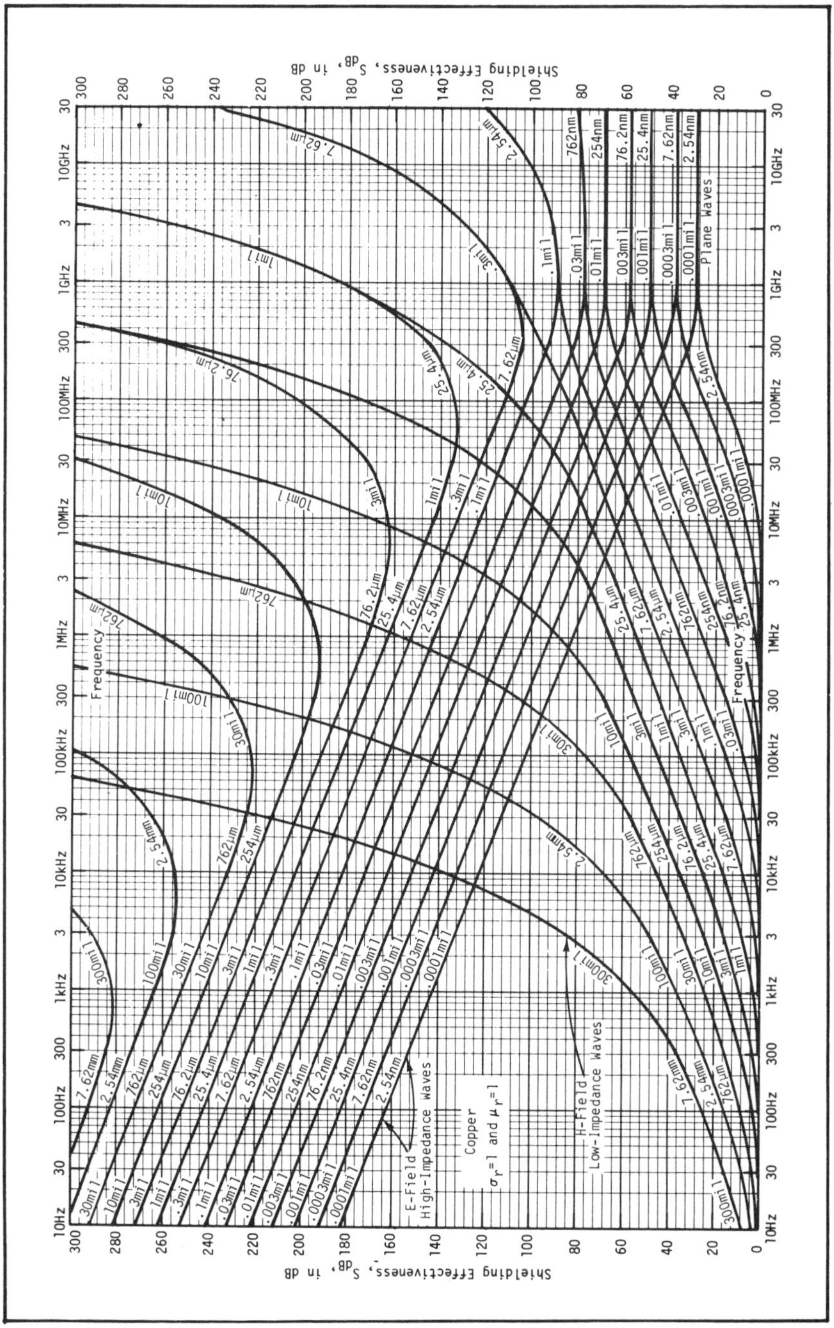

Figure A.1 - Shielding Effectiveness of Copper vs. Frequency for Source-to-Metal Distance of 10cm

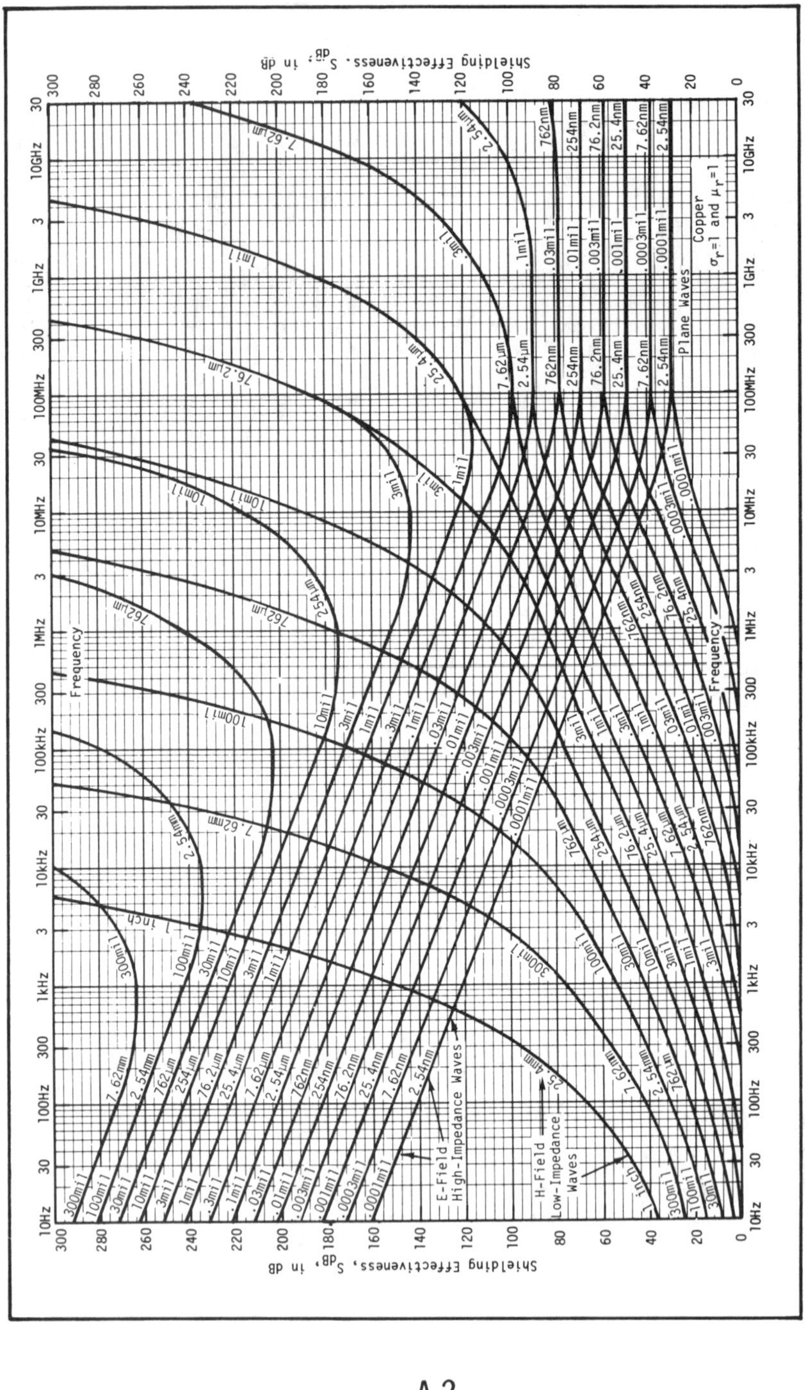

Figure A.2 - Shielding Effectiveness of Copper vs. Frequency for Source-to-Metal Distance of 1m

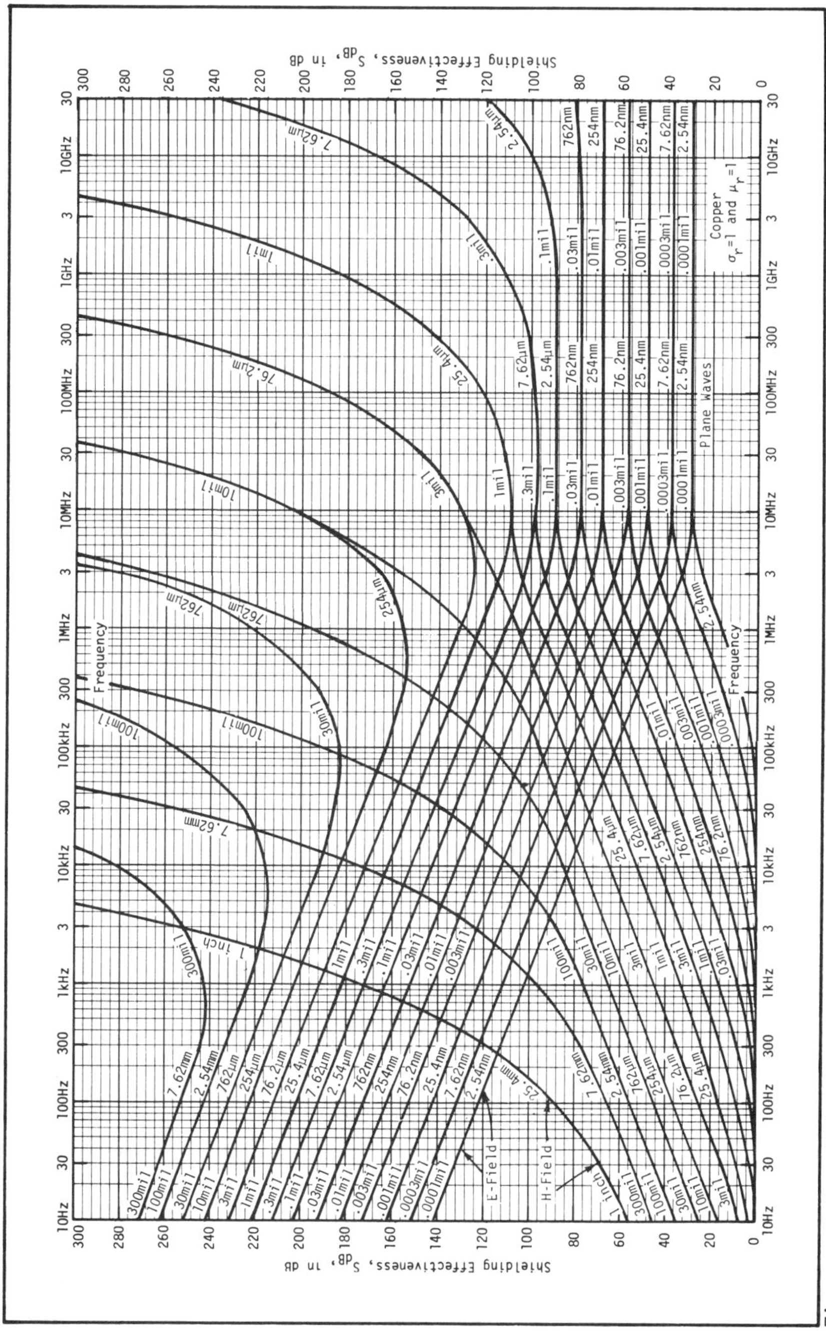

Figure A.3 - Shielding Effectiveness of Copper vs. Frequency for Source-to-Metal Distance of 10m

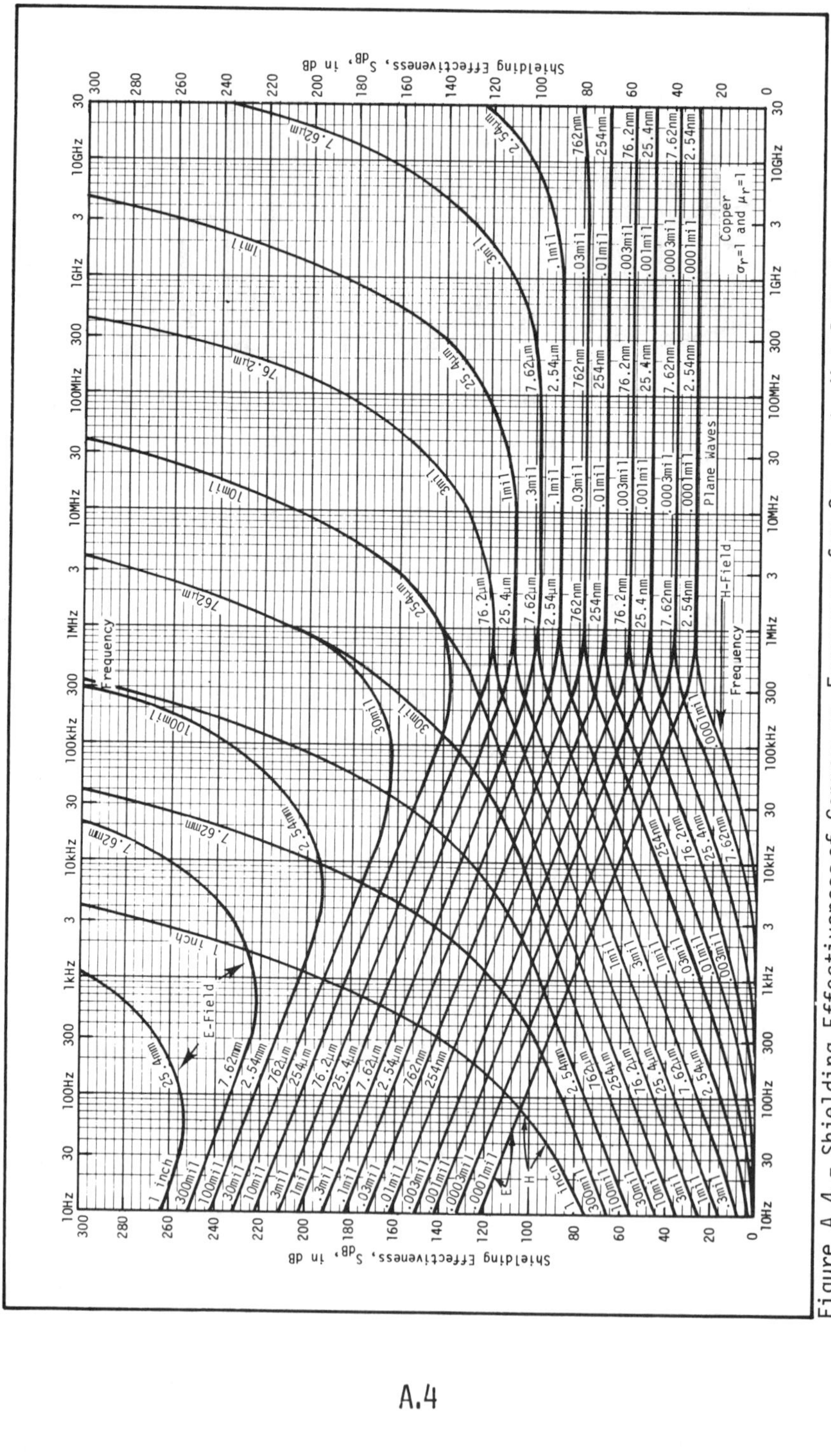

Figure A.4 - Shielding Effectiveness of Copper vs. Frequency for Source-to-Metal Distance of 100m

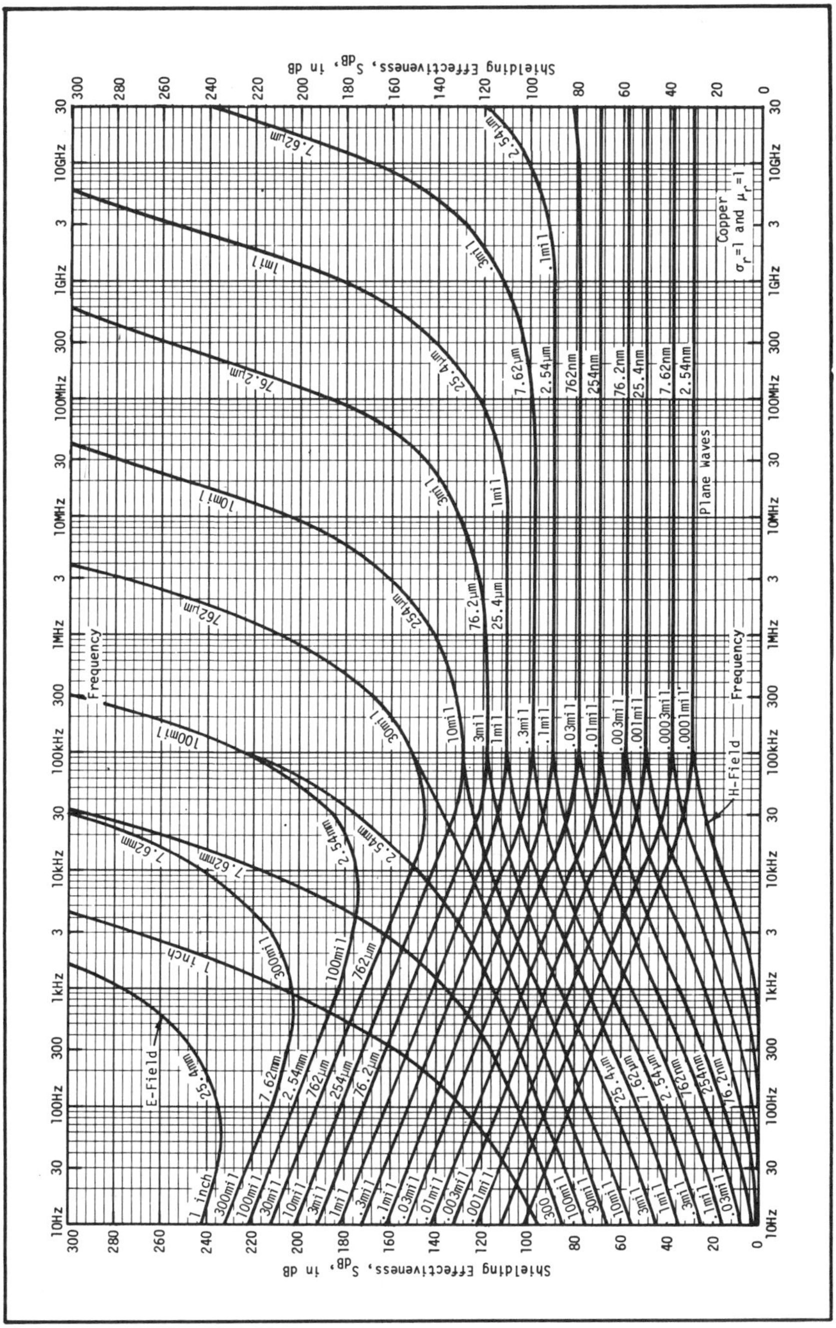

Figure A.5 - Shielding Effectiveness of Copper vs. Frequency for Source-to-Metal Distance of 1km

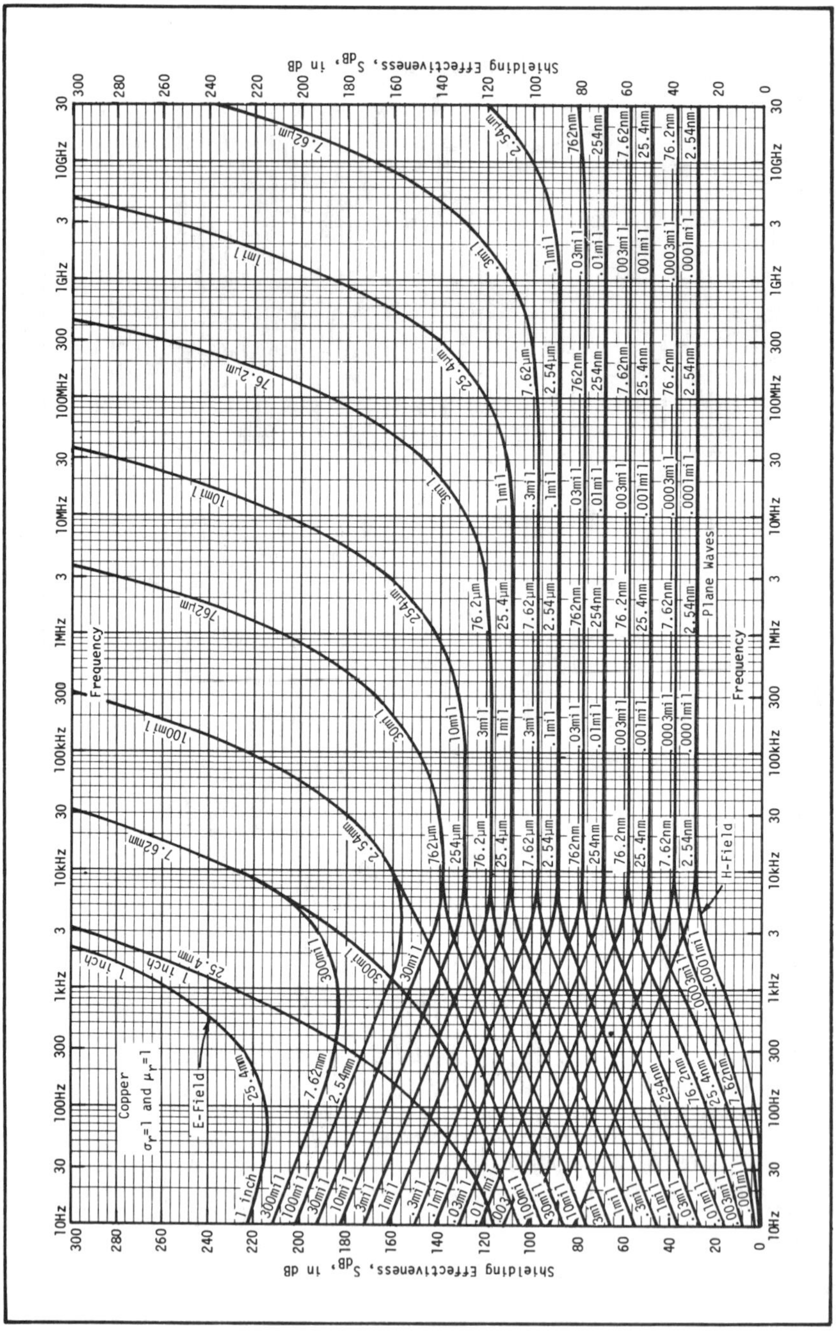

Figure A.6 - Shielding Effectiveness of Copper vs. Frequency for Source-to-Metal Distance of 10km

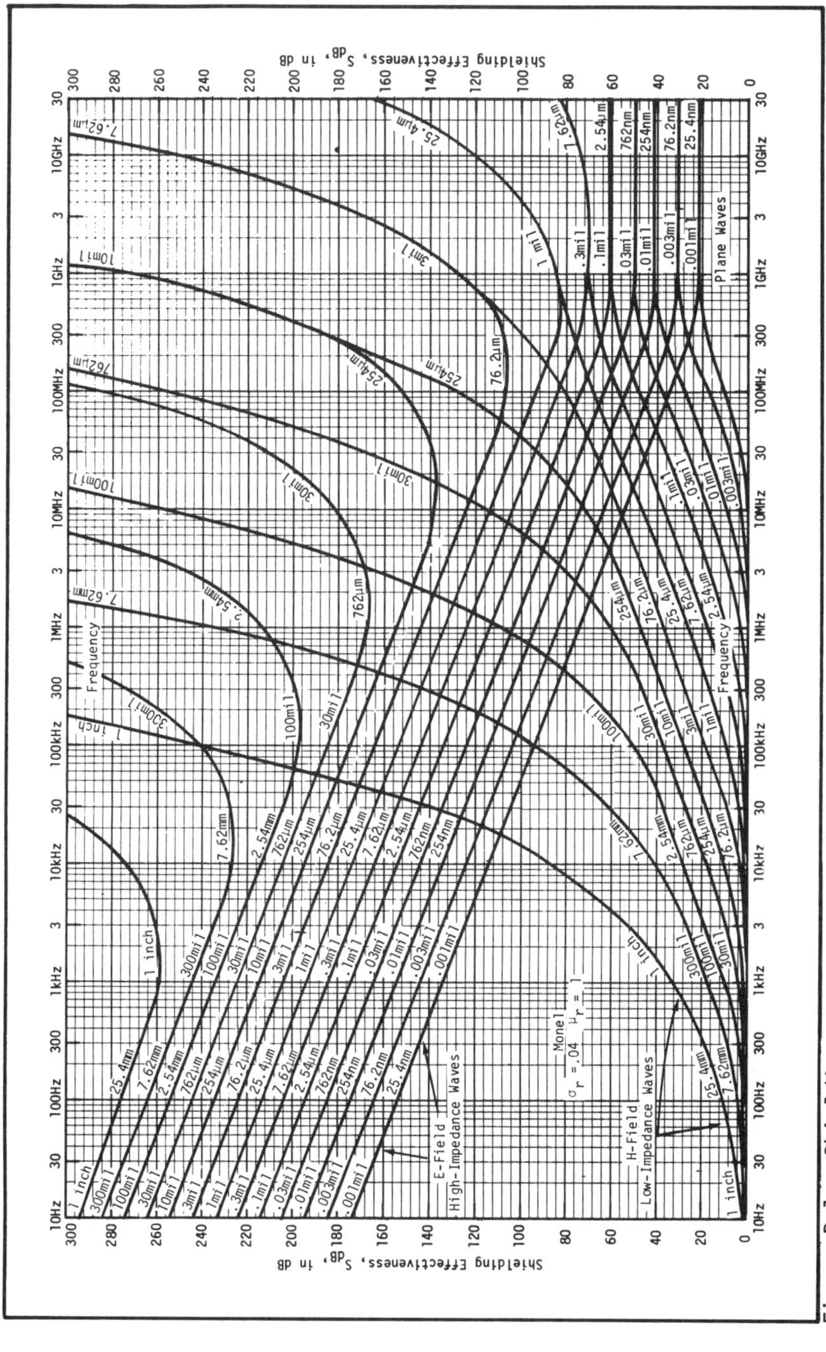

Figure B.1 - Shielding Effectiveness of Monel vs. Frequency for Source-to-Metal Distance of 10cm

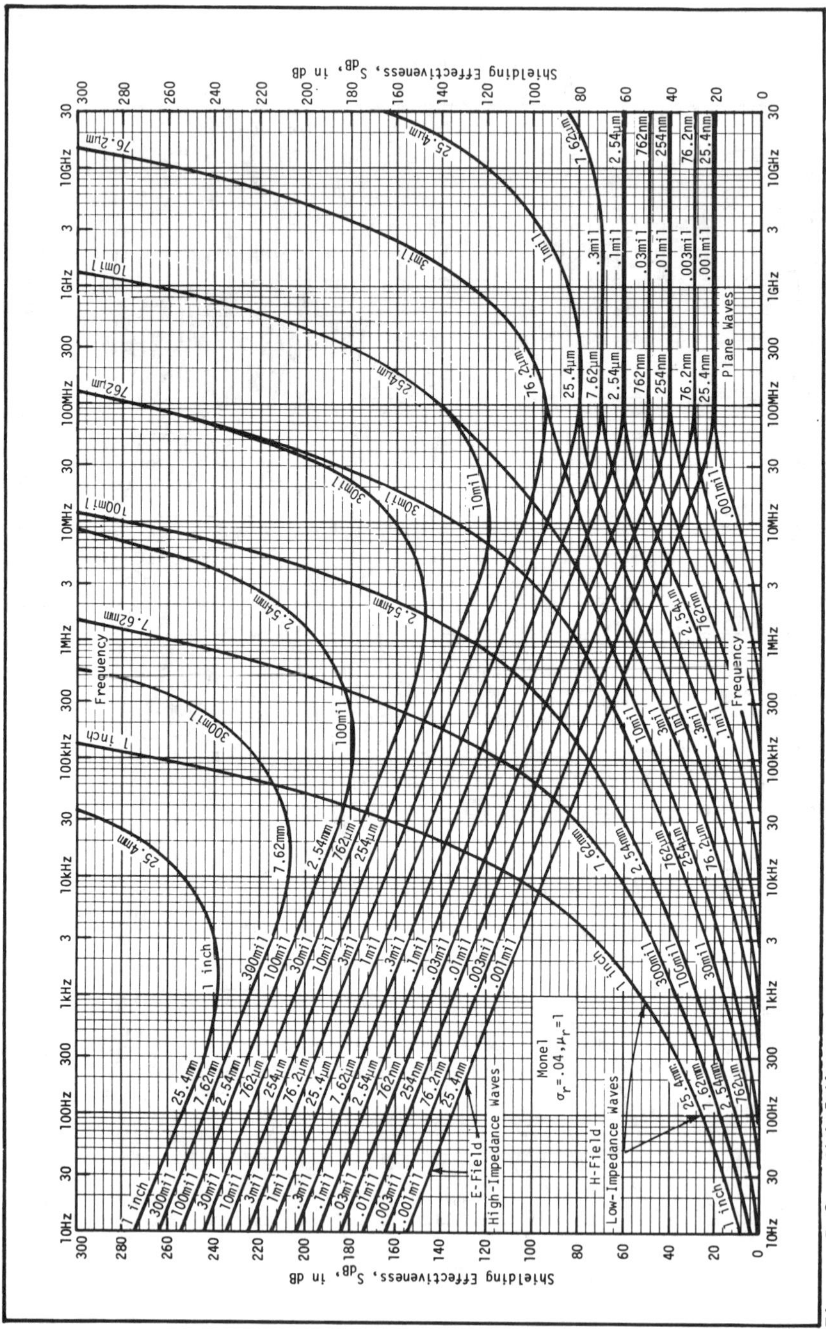

Figure B.2 - Shielding Effectiveness of Monel vs. Frequency for Source-to-Metal Distance of 1m

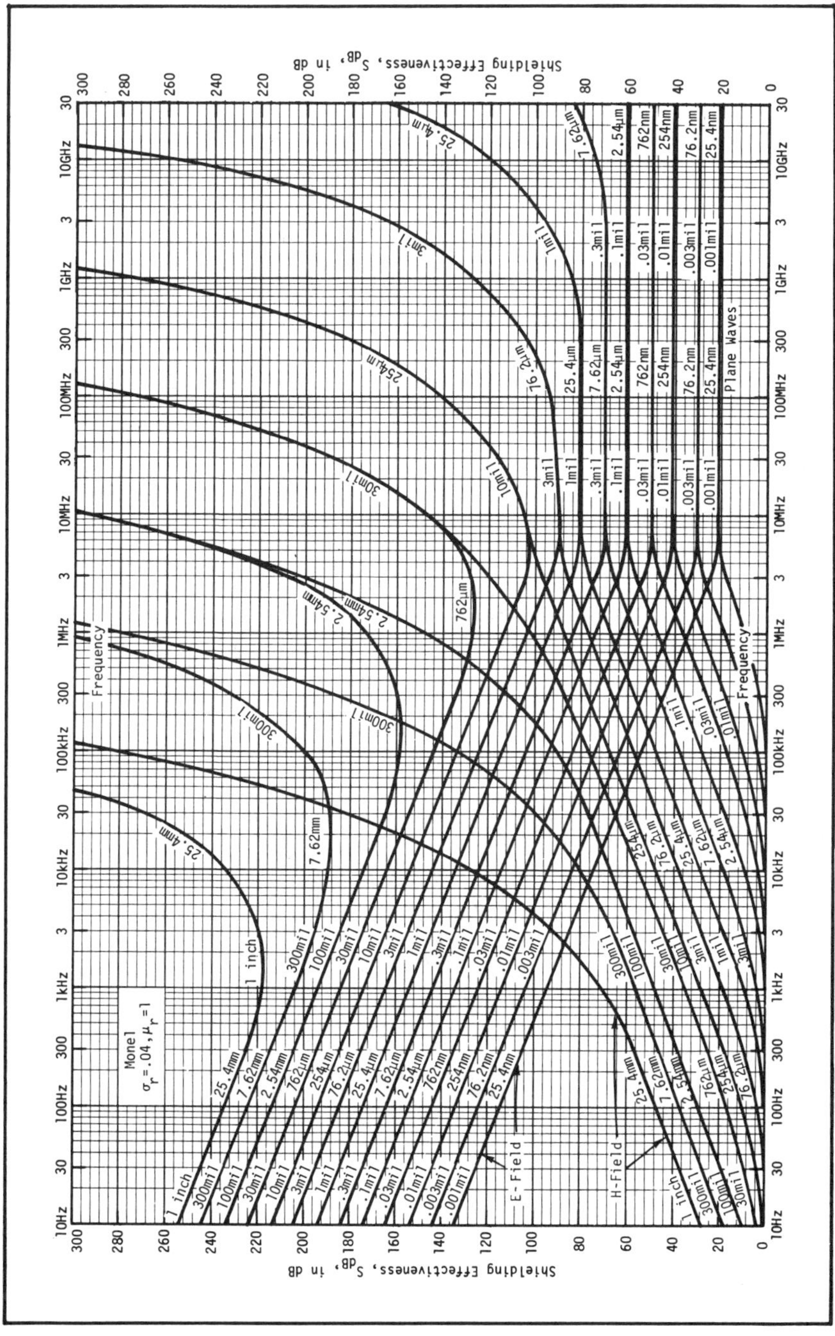

Figure B.3 - Shielding Effectiveness of Monel vs. Frequency for Source-to-Metal Distance of 10m

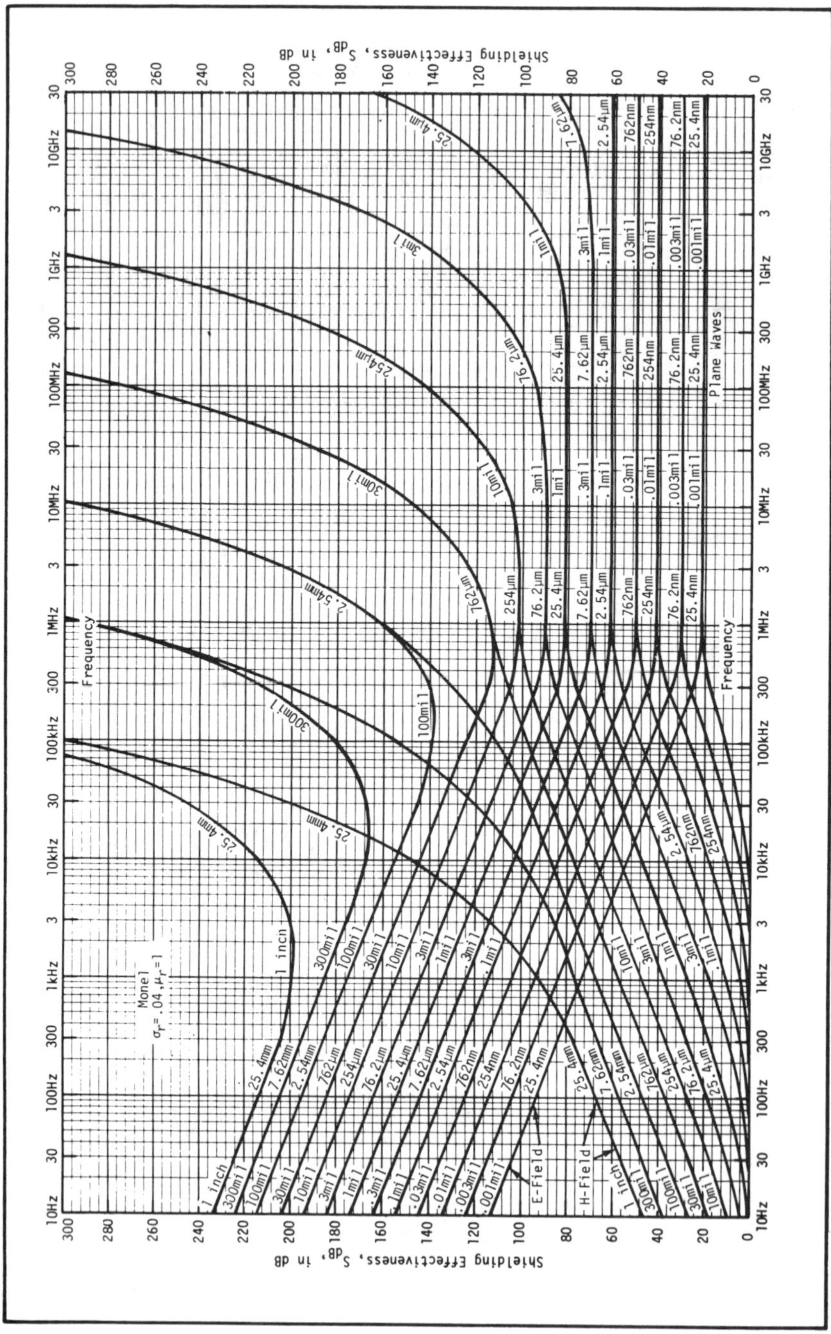

Figure B.4 - Shielding Effectiveness of Monel vs. Frequency for Source-to-Metal Distance of 100m

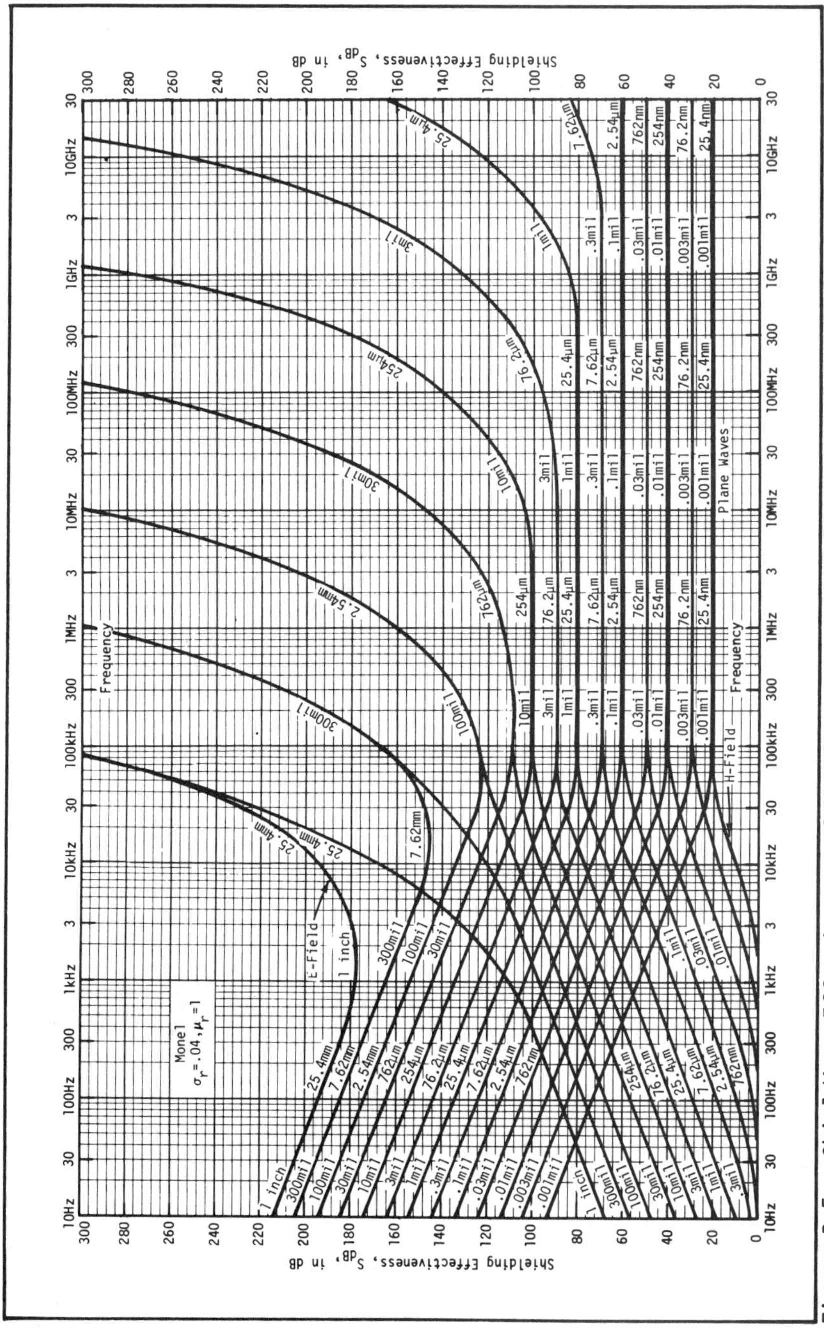

Figure B.5 - Shielding Effectiveness of Monel vs. Frequency for Source-to-Metal Distance of 1km

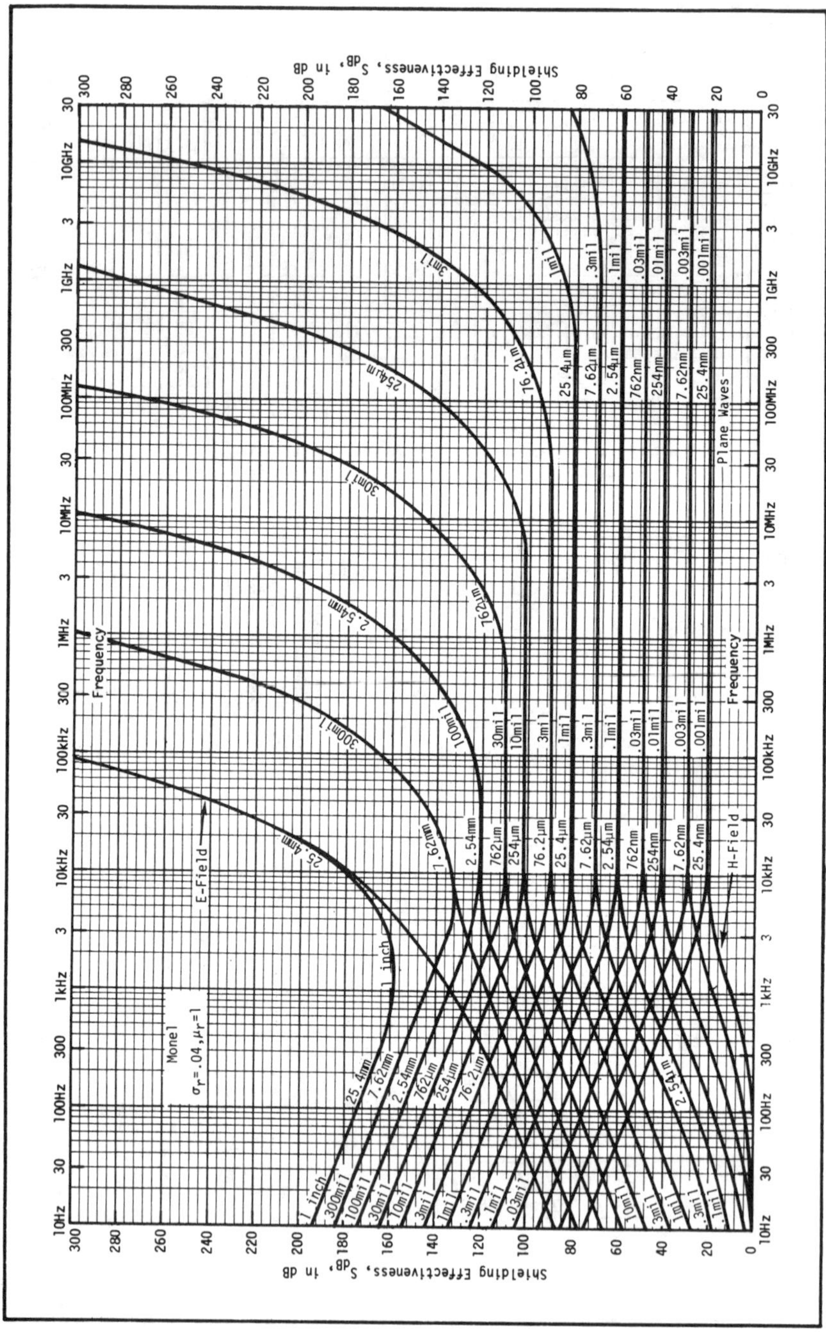

Figure B.6 - Shielding Effectiveness of Monel vs. Frequency for Source-to-Metal Distance of 10km

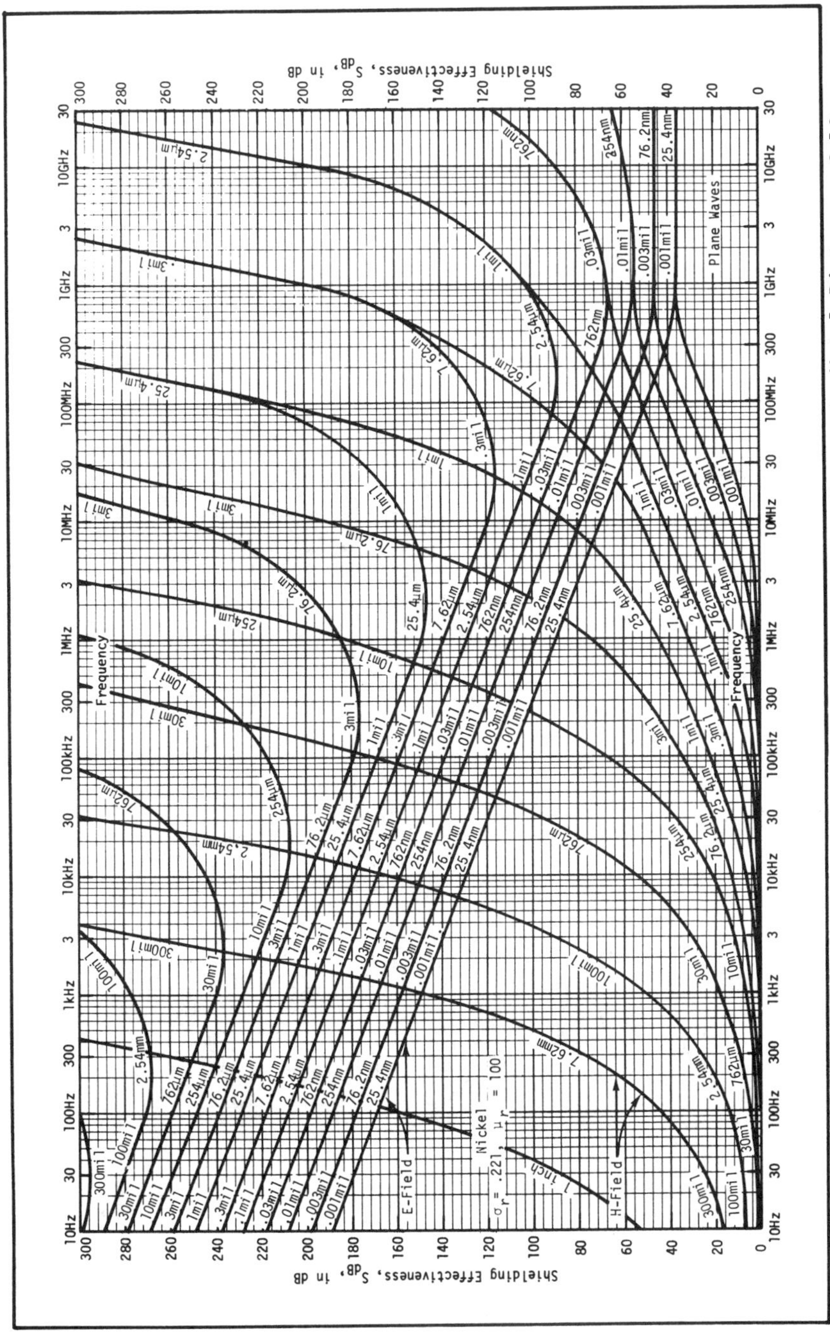

Figure C.1 - Shielding Effectiveness of Nickel vs. Frequency for Source-to-Metal Distance of 10cm

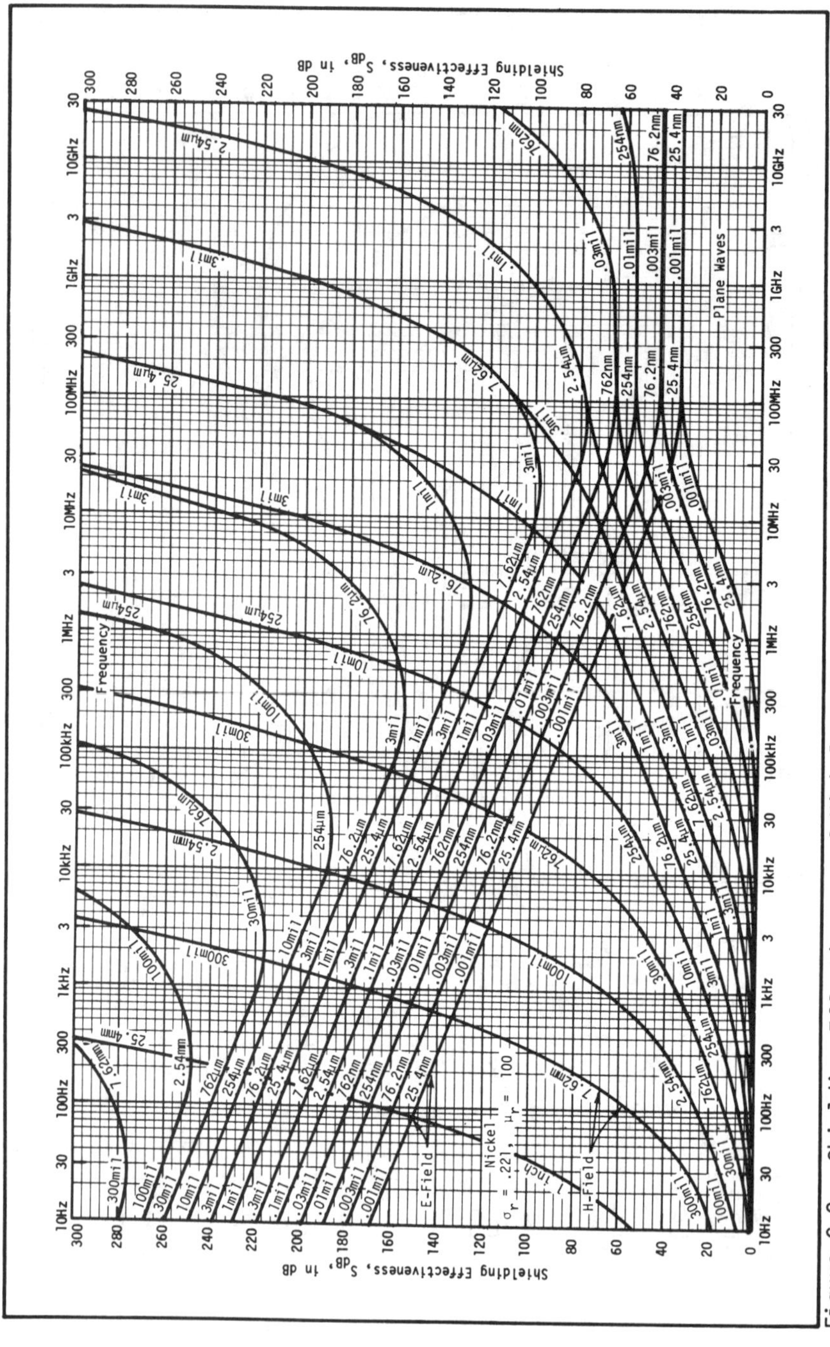

Figure C.2 - Shielding Effectiveness of Nickel vs. Frequency for Source-to-Metal Distance of 1m

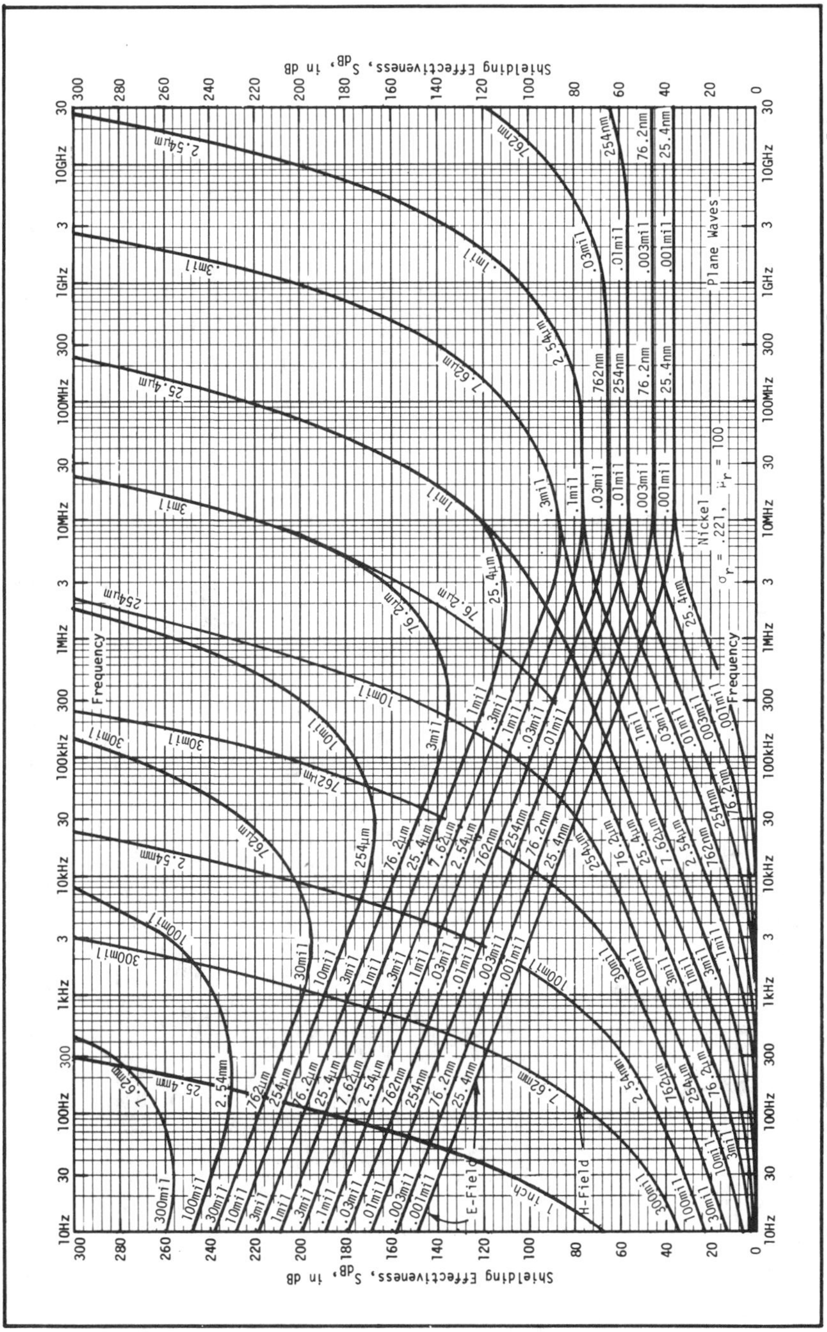

Figure C.3 - Shielding Effectiveness of Nickel vs. Frequency for Source-to-Metal Distance of 10m

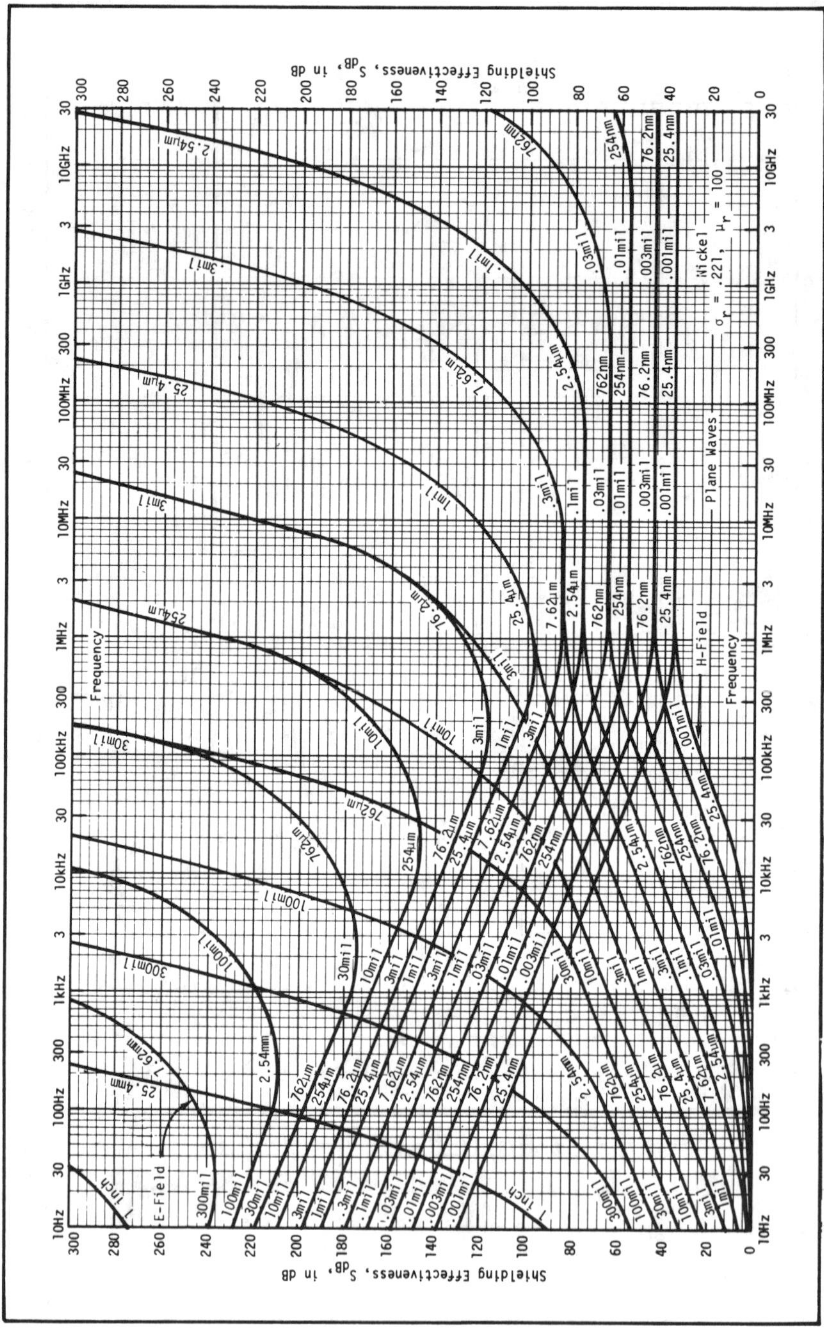

Figure C.4 - Shielding Effectiveness of Nickel vs. Frequency for Source-to-Metal Distance of 100m

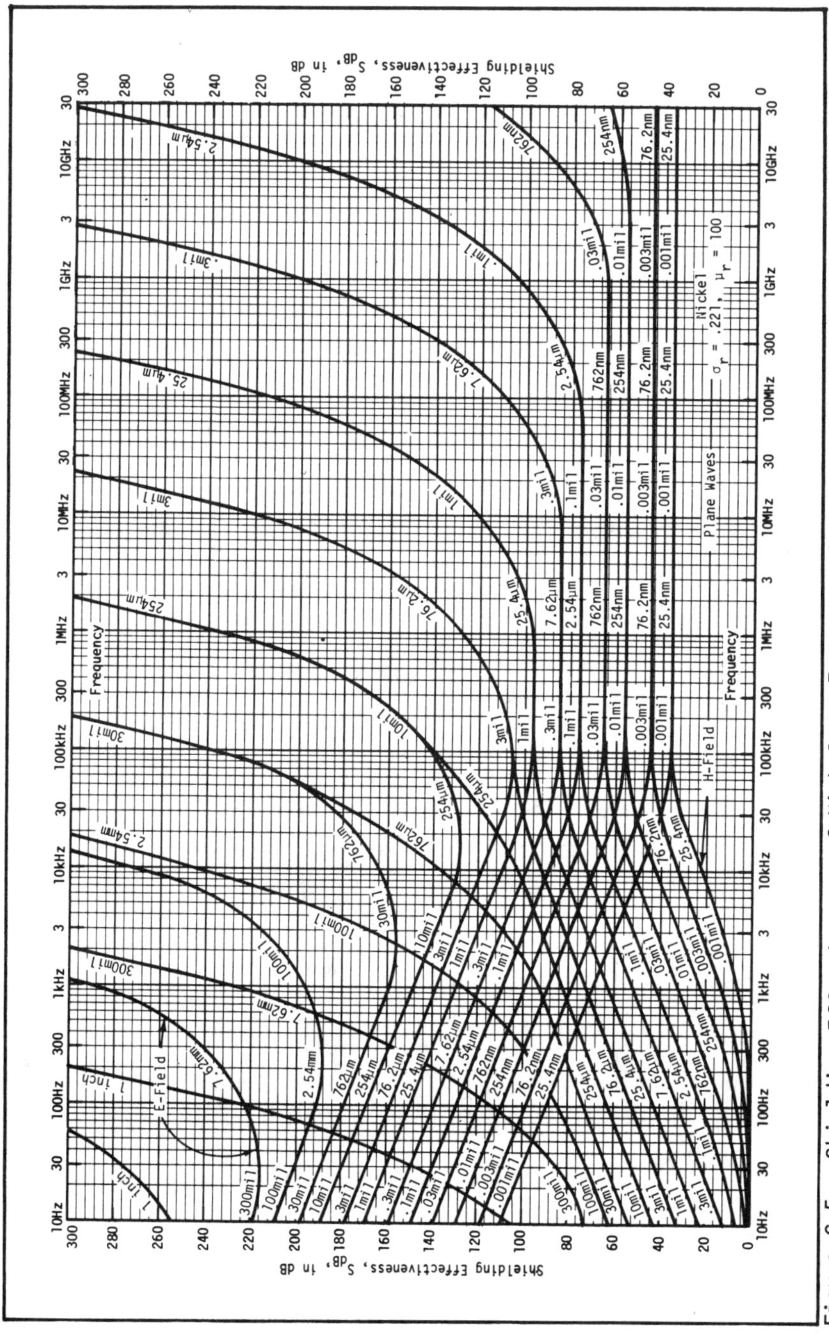

Figure C.5 - Shielding Effectiveness of Nickel vs. Frequency for Source-to-Metal Distance of 1km

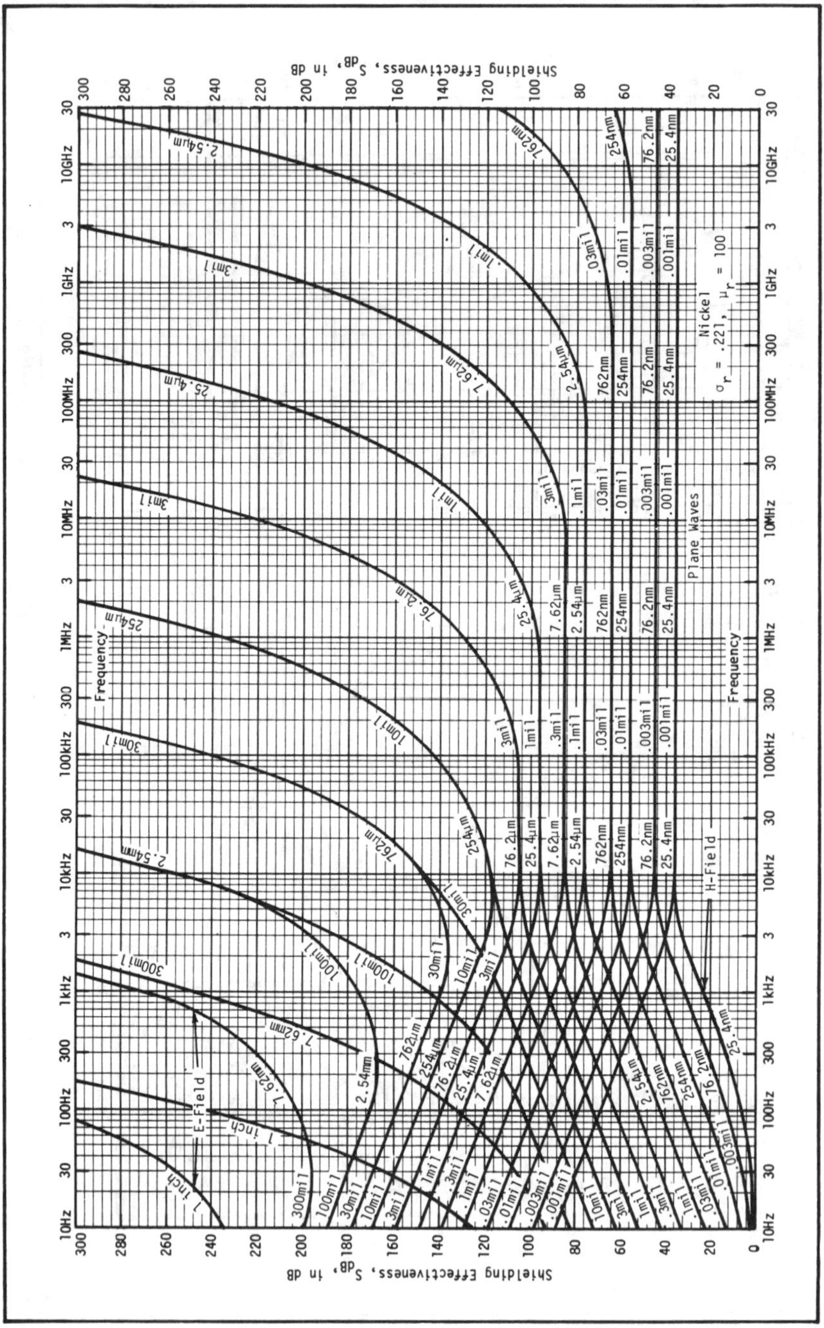

Figure C.6 - Shielding Effectiveness of Nickel vs. Frequency for Source-to-Metal Distance of 10km

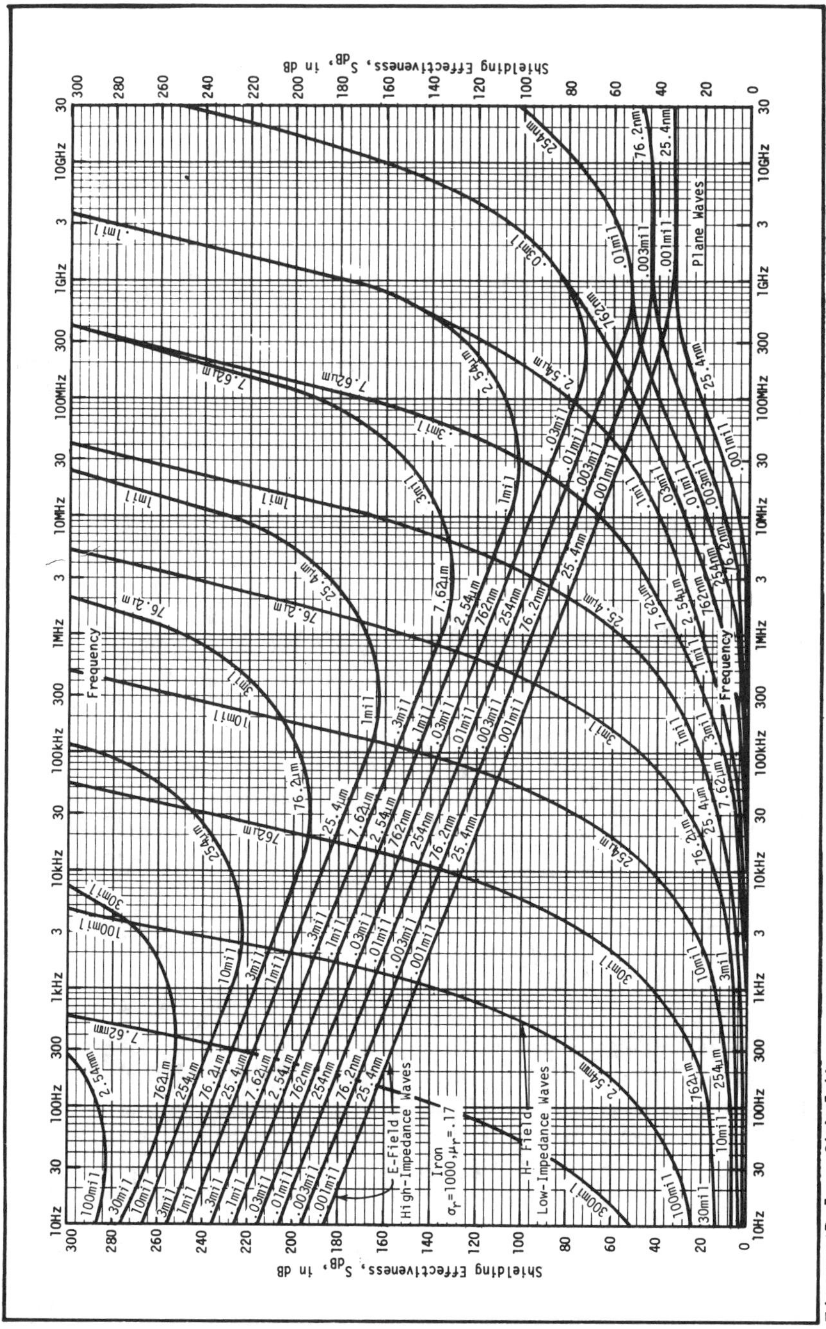

Figure D.1 - Shielding Effectiveness of Iron vs. Frequency for Source-to-Metal Distance of 10cm

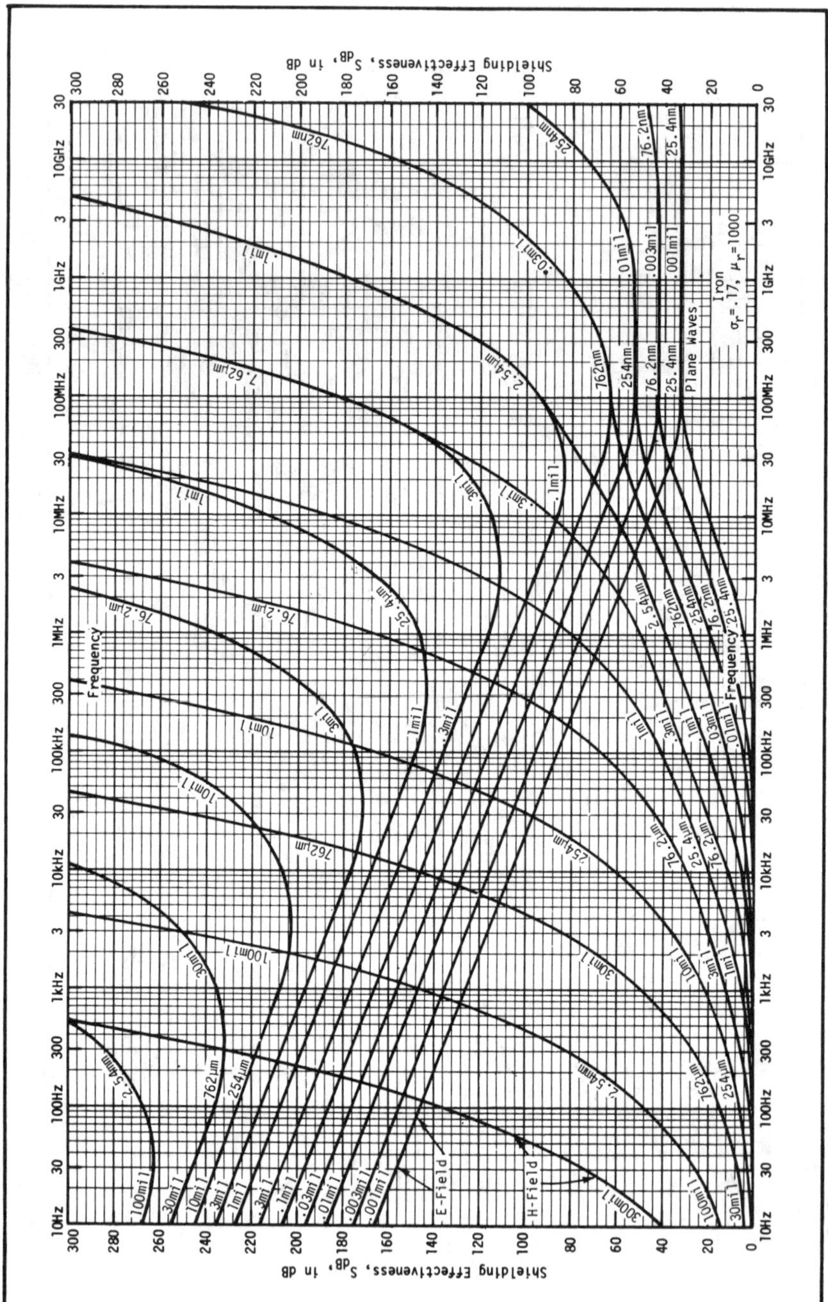

Figure D.2 - Shielding Effectiveness of Iron vs. Frequency for Source-to-Metal Distance of 1m

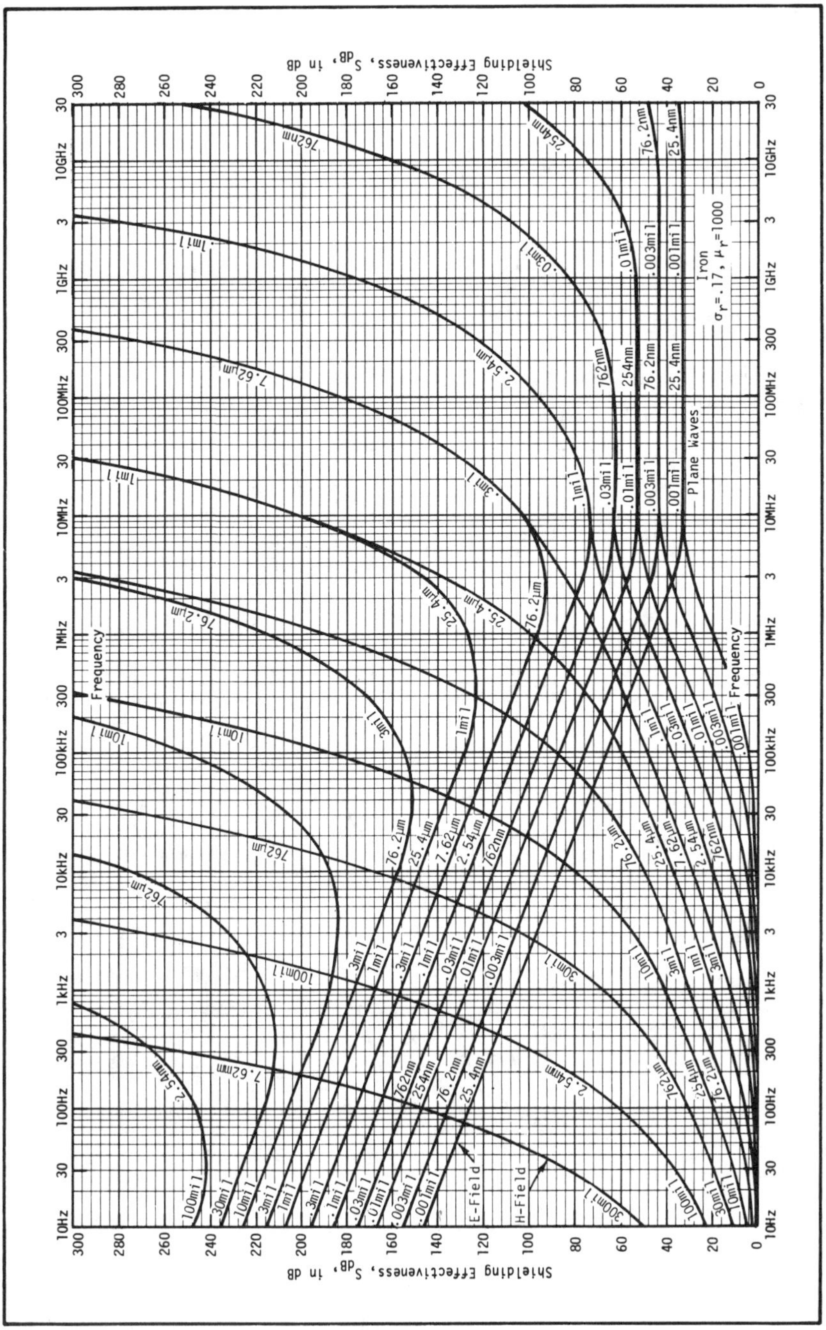

Figure D.3 - Shielding Effectiveness of Iron vs. Frequency for Source-to-Metal Distance of 10m

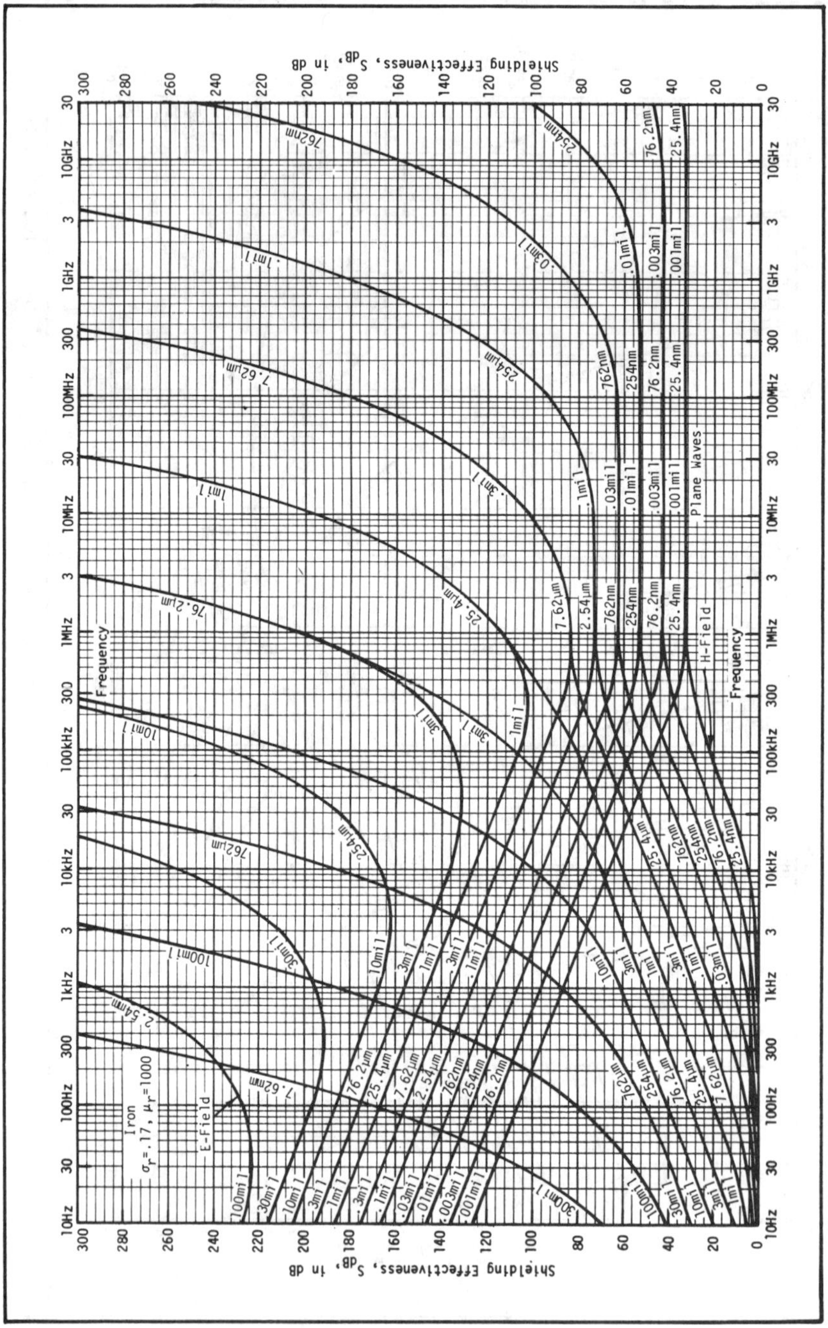

Figure D.4 - Shielding Effectiveness of Iron vs. Frequency for Source-to-Metal Distance of 100m

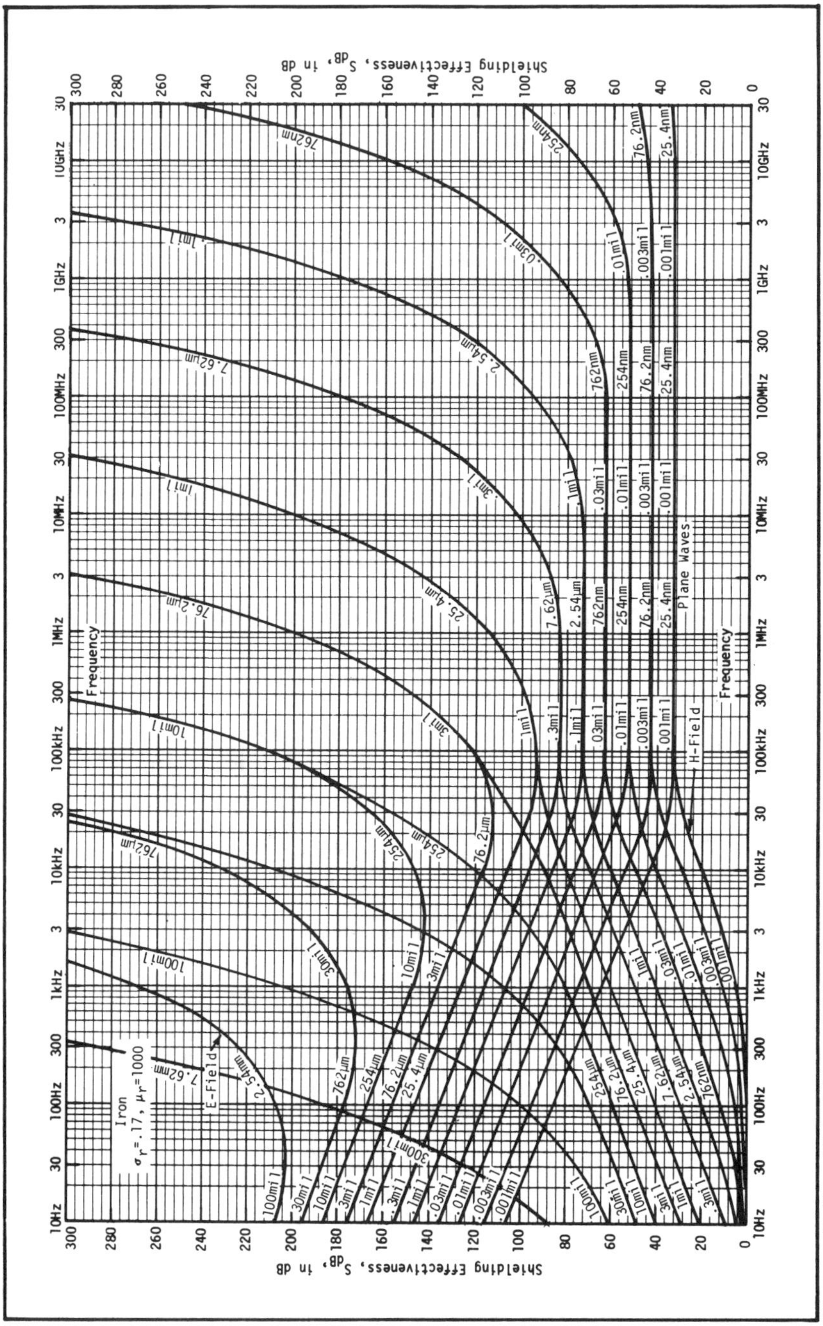

Figure D.5 - Shielding Effectiveness of Iron vs. Frequency for Source-to-Metal Distance of 1km

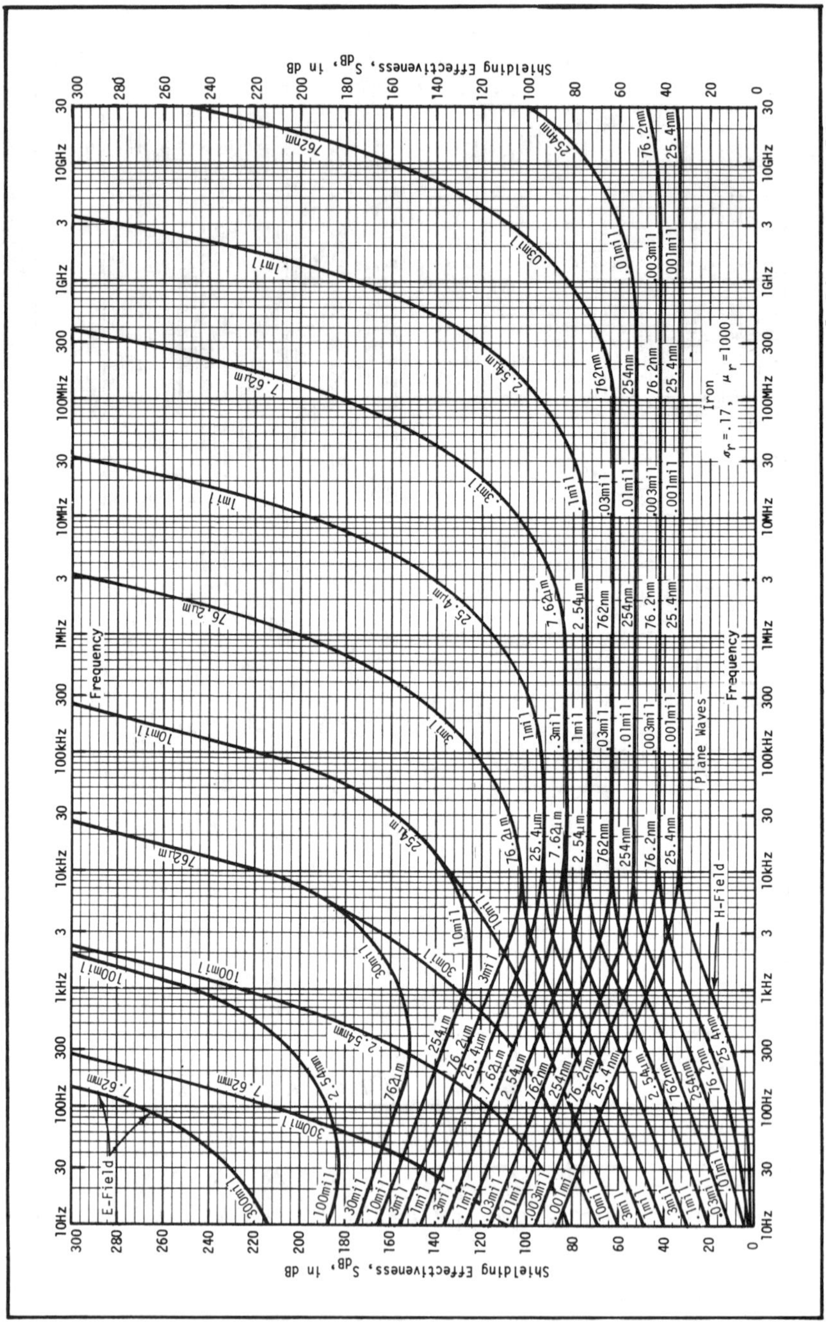

Figure D.6 - Shielding Effectiveness of Iron vs. Frequency for Source-to-Metal Distance of 10km

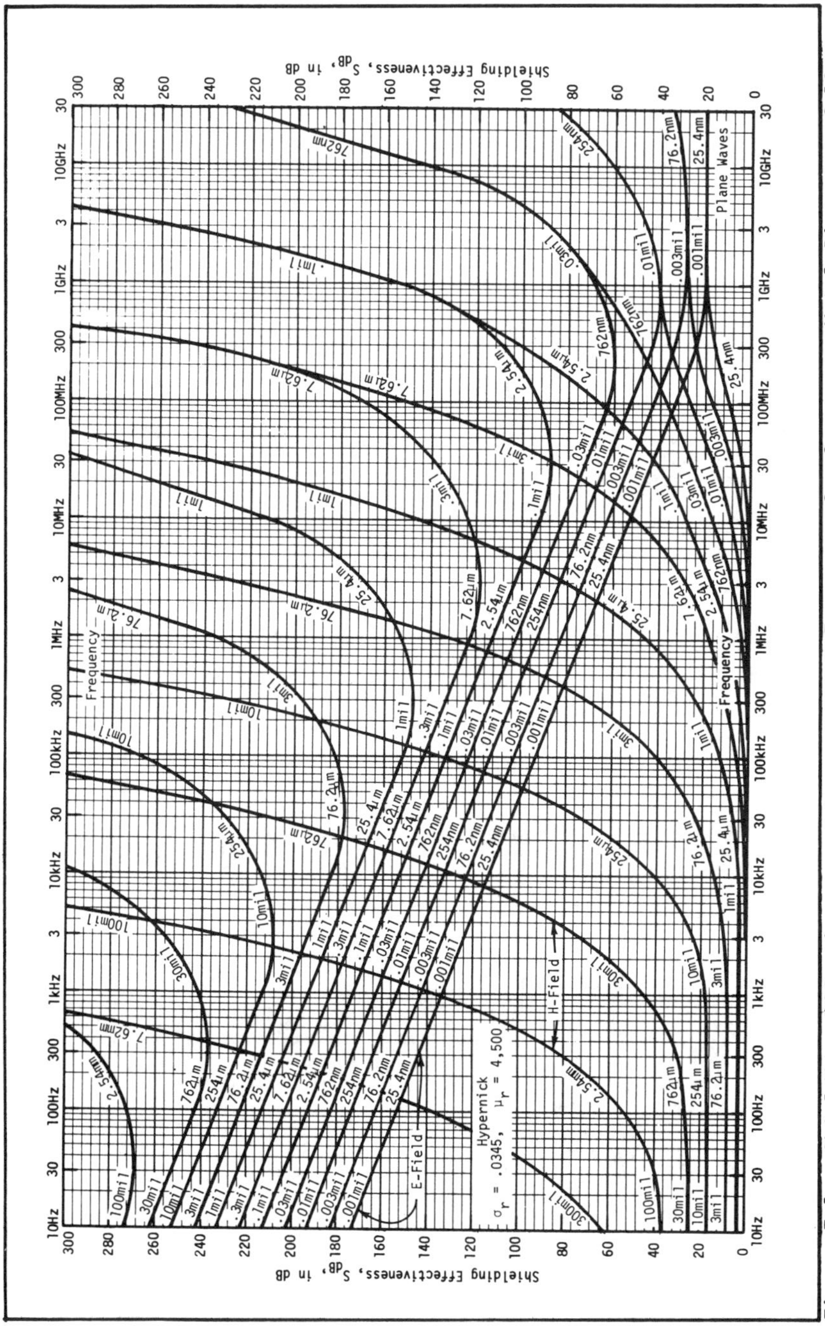

Figure E.1 - Shielding Effectiveness of Hypernick vs. Frequency for Source-to-Metal Distance of 10cm

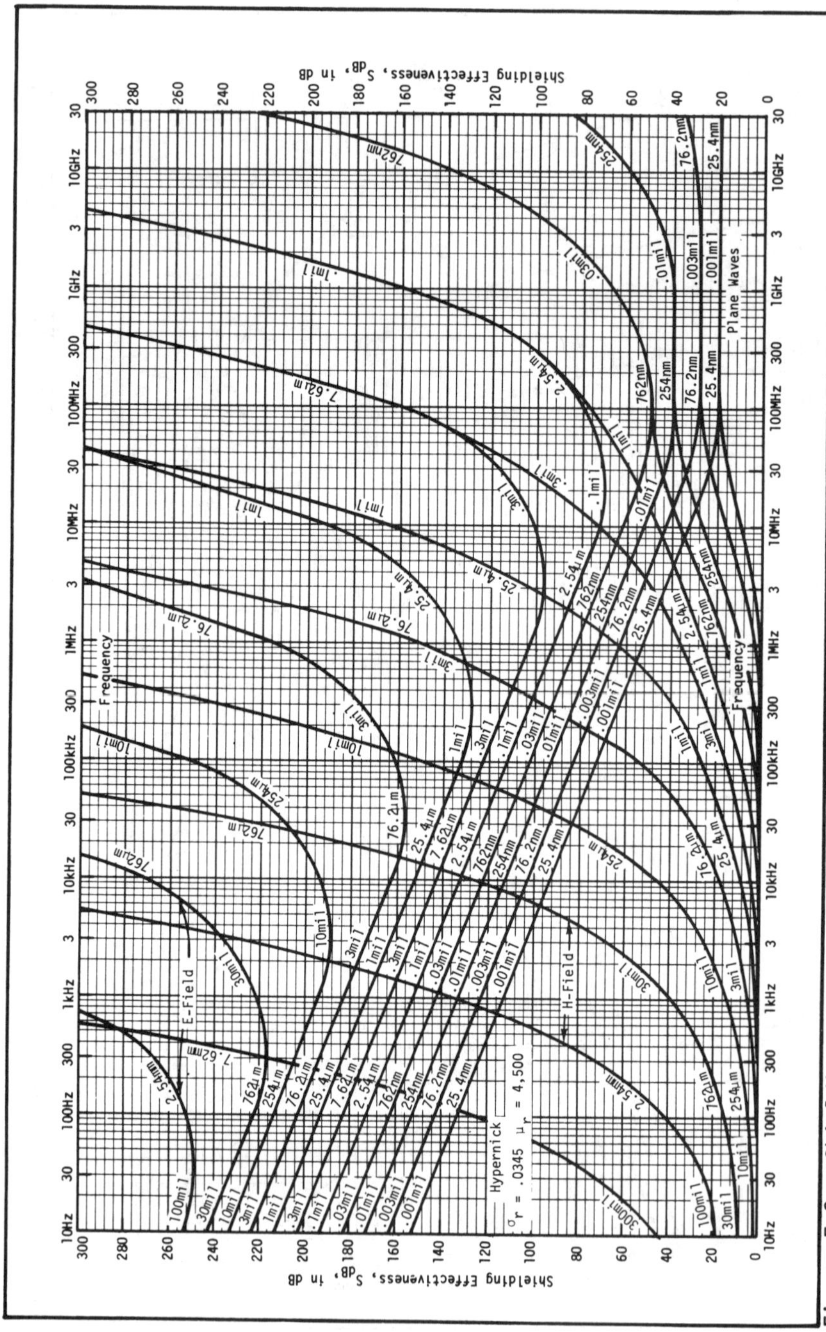

Figure E.2 - Shielding Effectiveness of Hypernick vs. Frequency for Source-to-Metal Distance of 1m

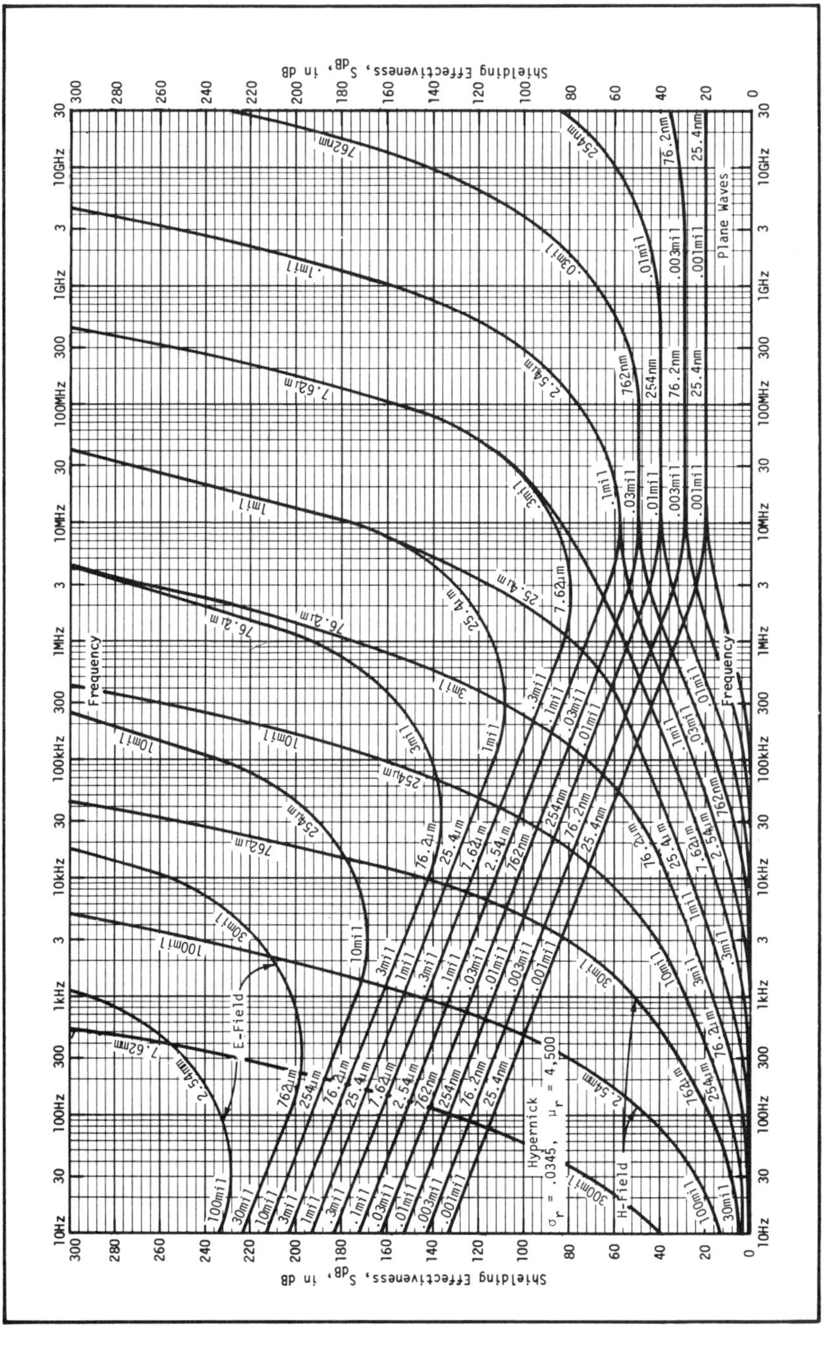

Figure E.3 - Shielding Effectiveness of Hypernick vs. Frequency for Source-to-Metal Distance of 10m

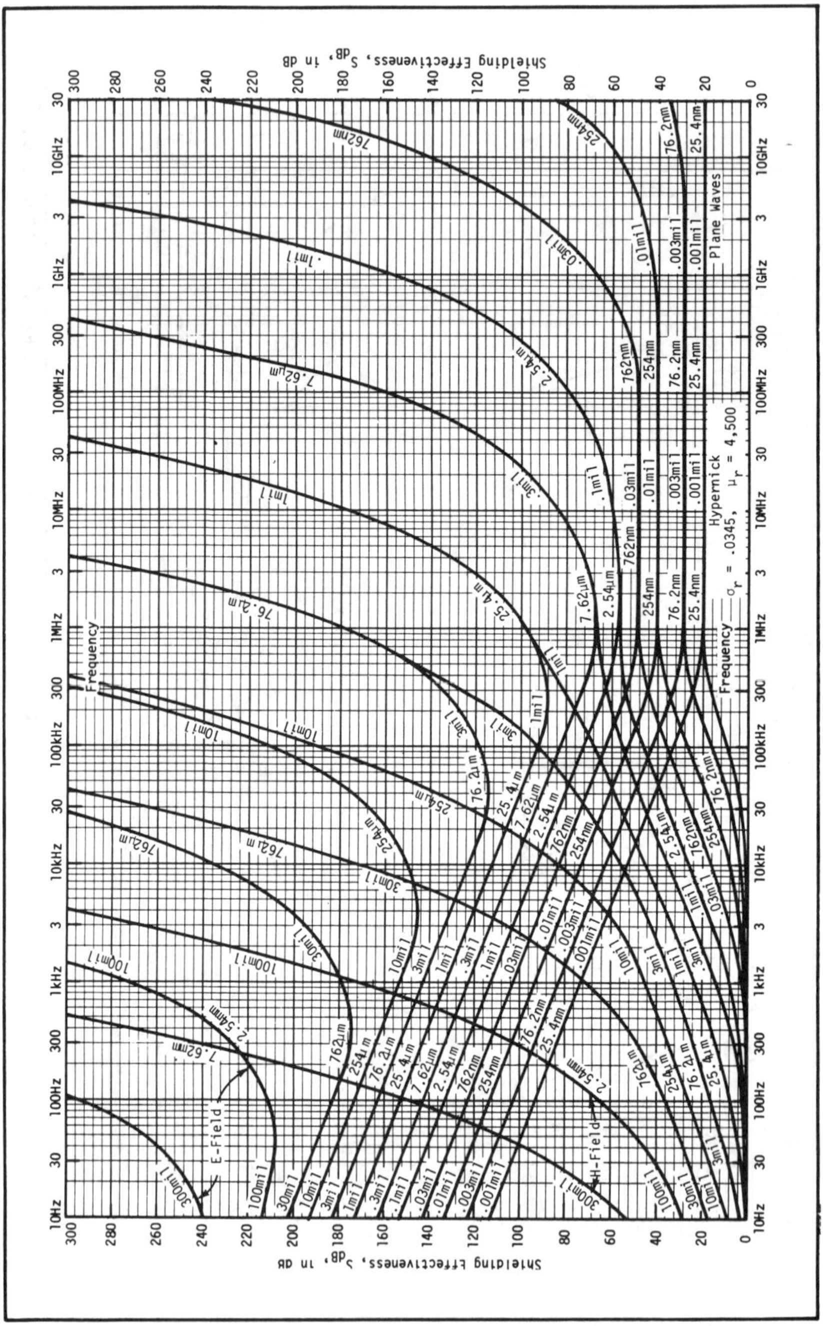

Figure E.4 - Shielding Effectiveness of Hypernick vs. Frequency for Source-to-Metal Distance of 100m

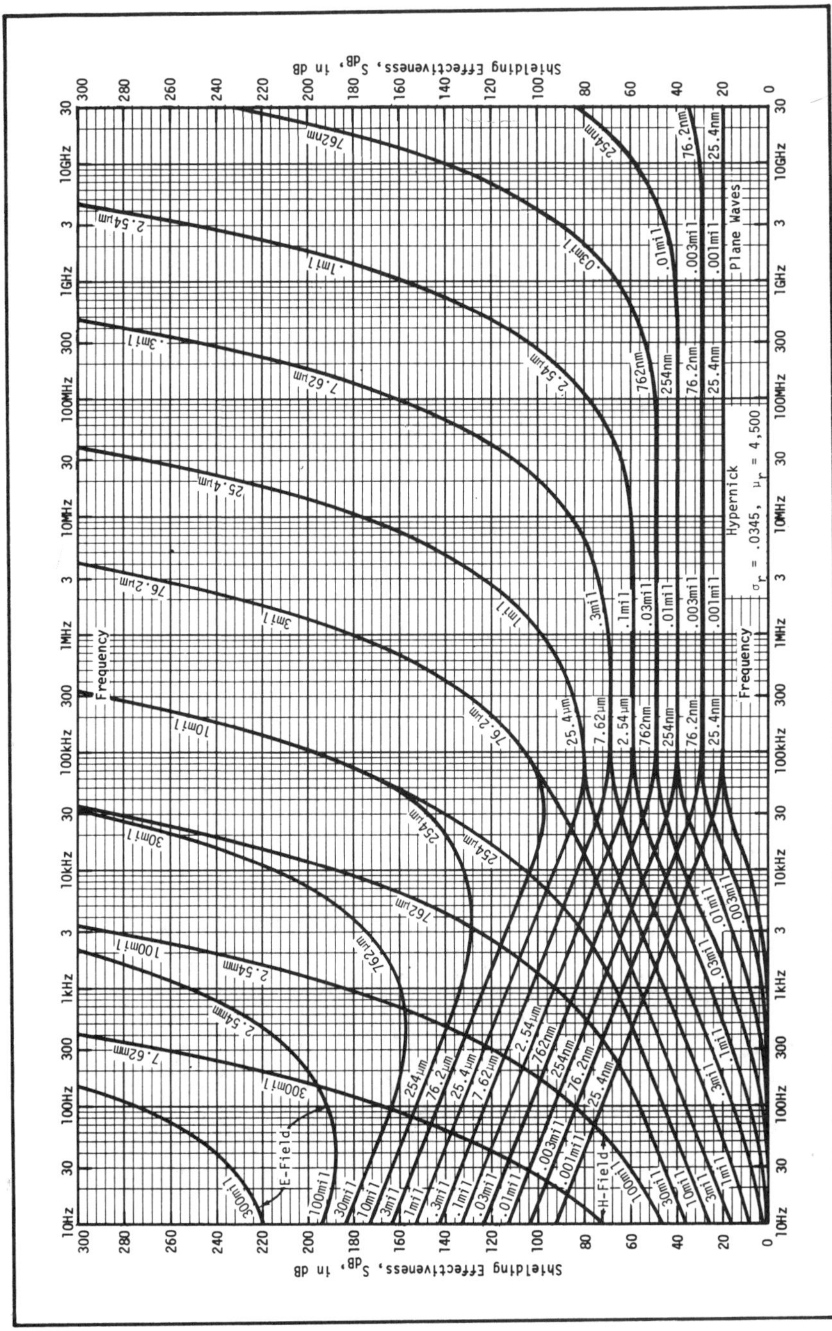

Figure E.5 - Shielding Effectiveness of Hypernick vs. Frequency for Source-to-Metal Distance of 1km

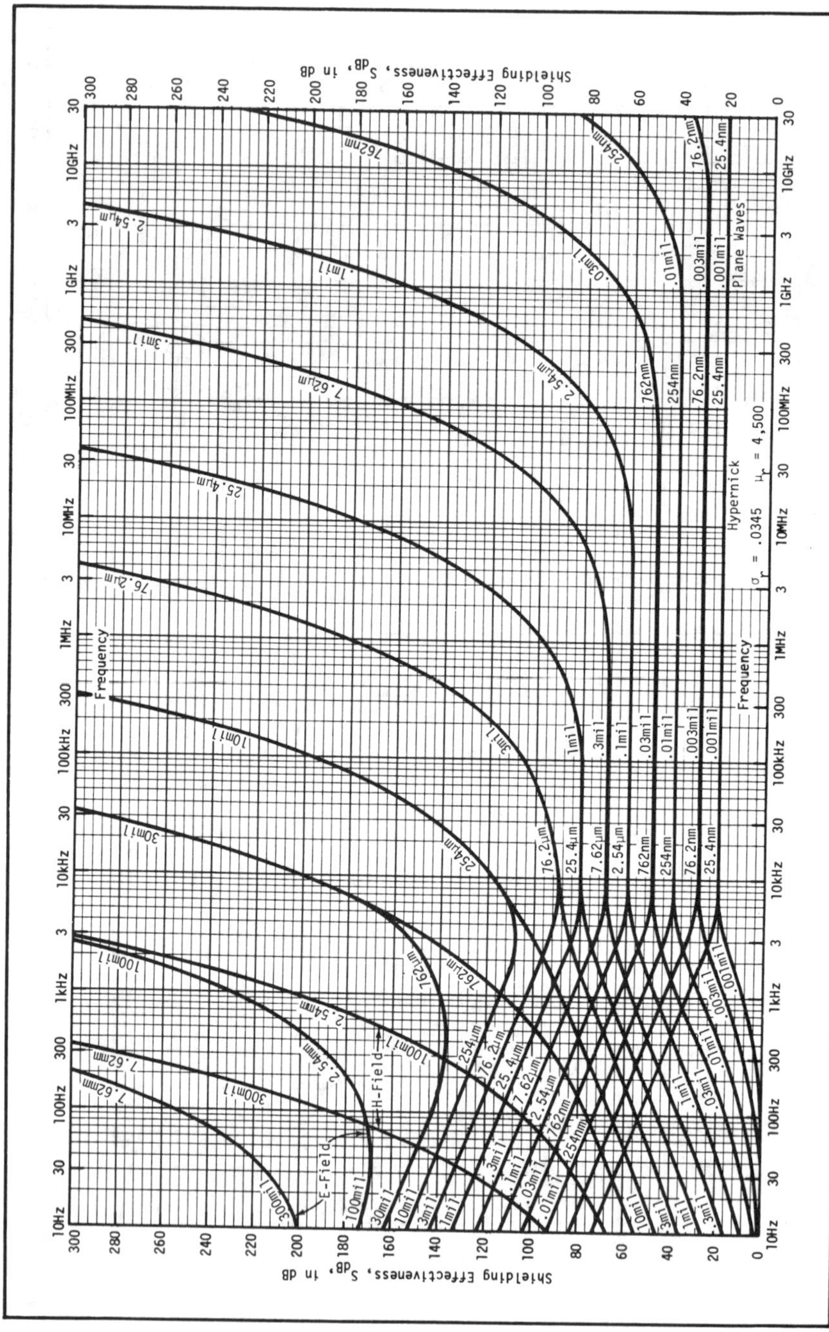

Figure E.6 - Shielding Effectiveness of Hypernick vs. Frequency for Source-to-Metal Distance of 10km

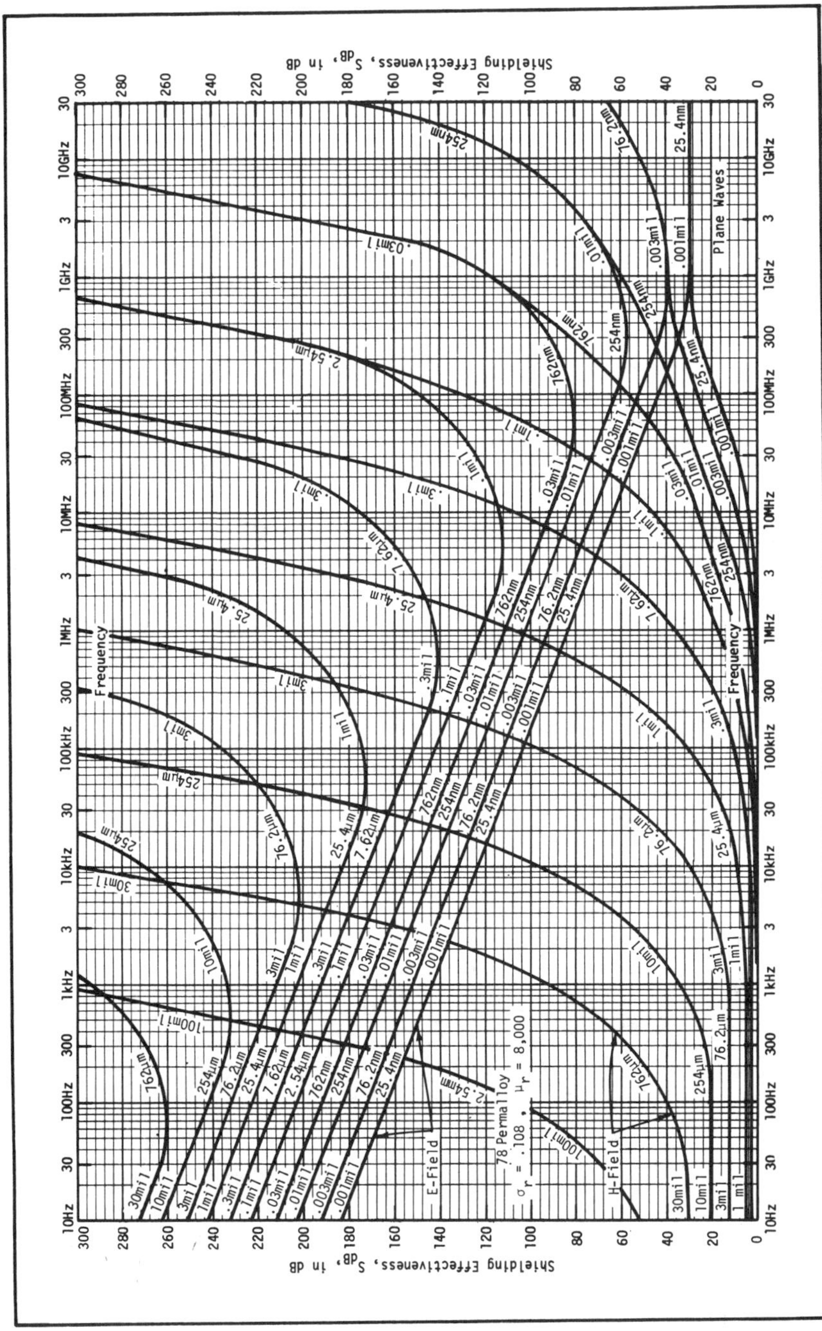

Figure F.1 - Shielding Effectiveness of 78 Permalloy vs. Frequency for Source-to-Metal Distance of 10cm

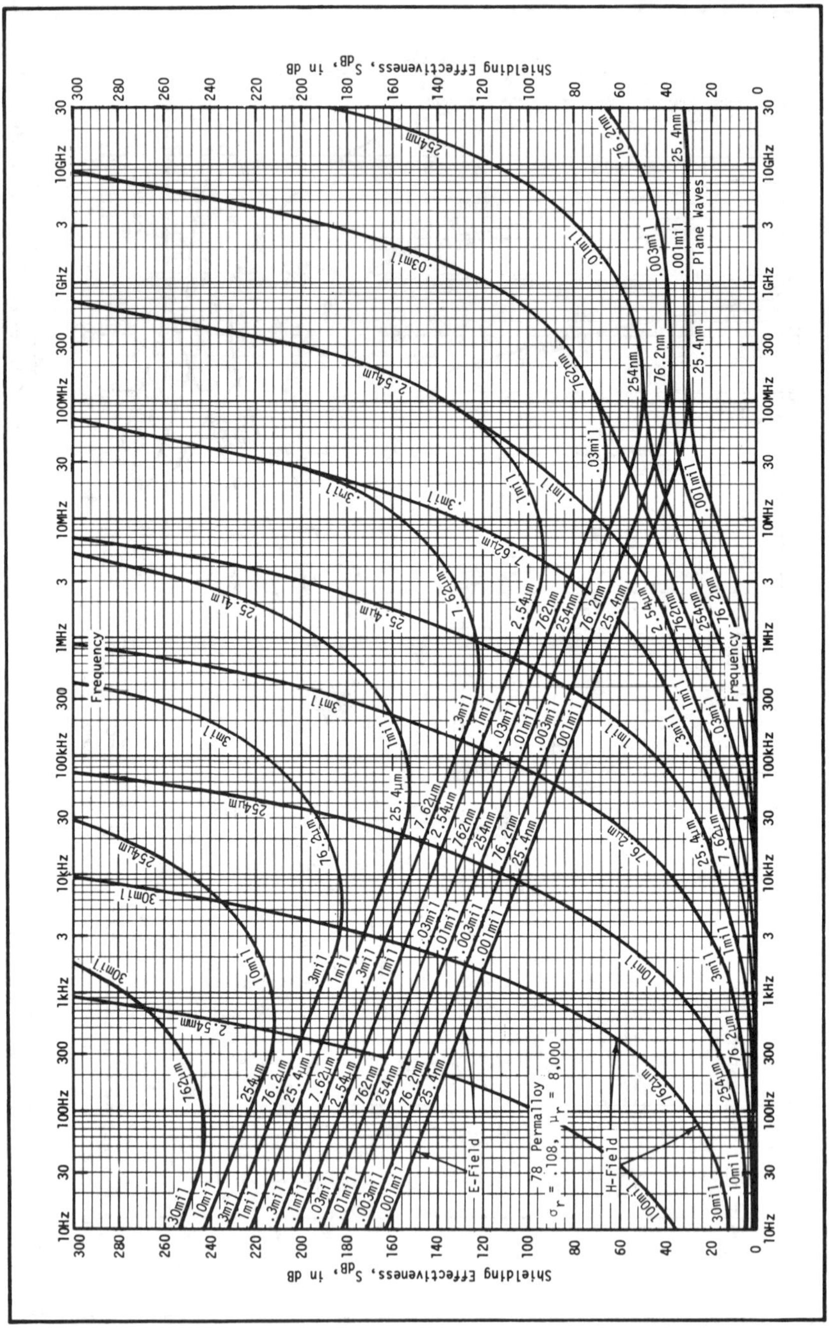

Figure F.2 - Shielding Effectiveness of 78 Permalloy vs. Frequency for Source-to-Metal Distance of 1m

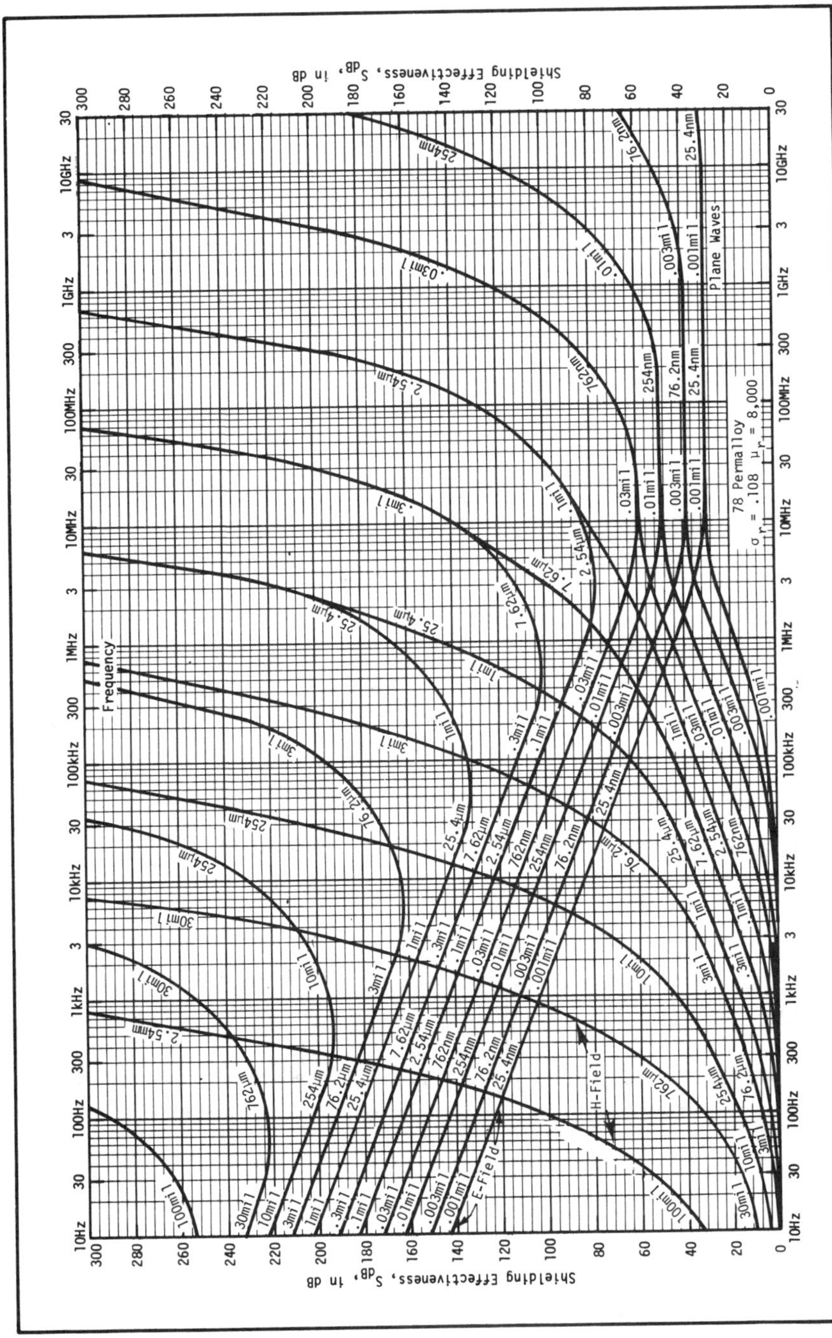

Figure F.3 - Shielding Effectiveness of 78 Permalloy vs. Frequency for Source-to-Metal Distance of 10m

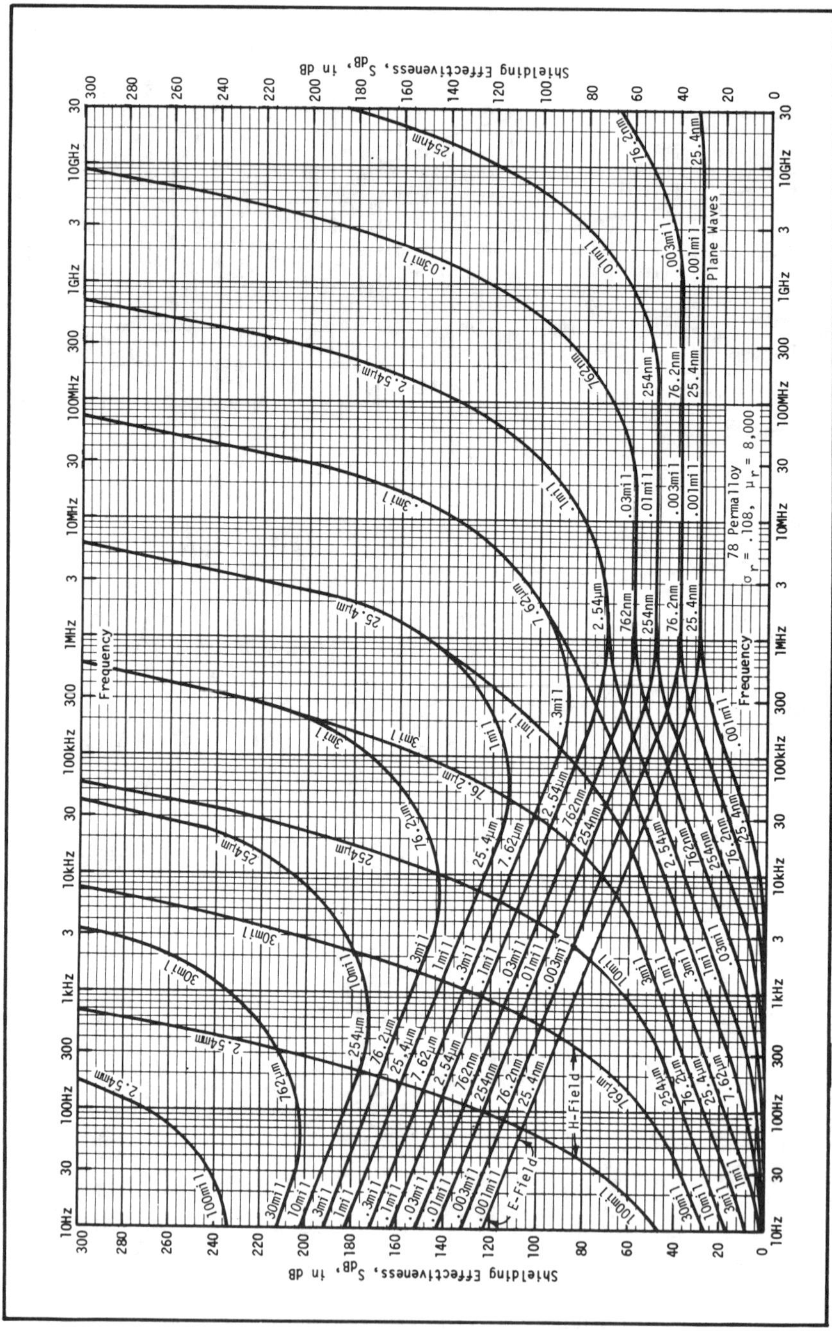

Figure F.4 - Shielding Effectiveness of 78 Permalloy vs. Frequency for Source-to-Metal Distance of 100m

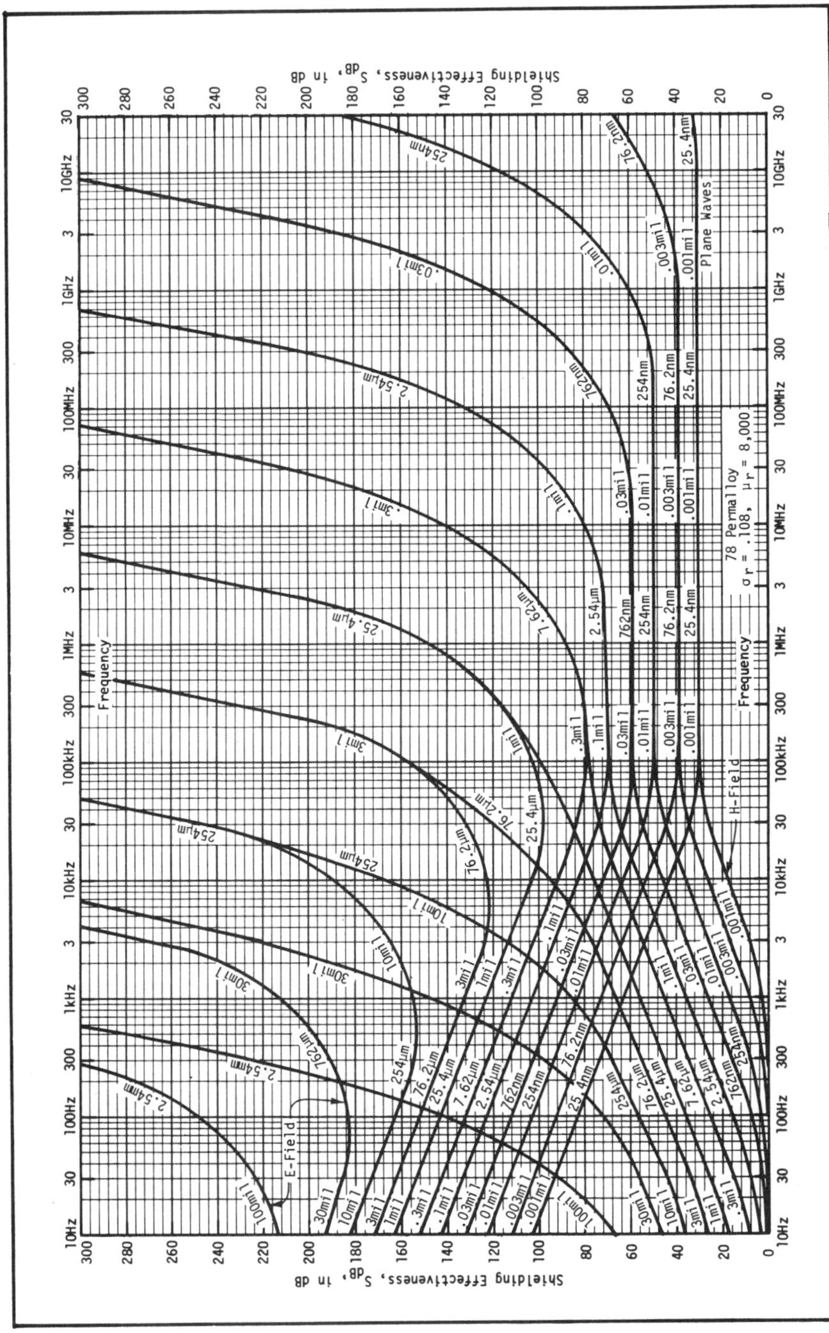

Figure F.5 - Shielding Effectiveness of 78 Permalloy vs. Frequency for Source-to-Metal Distance of 1km

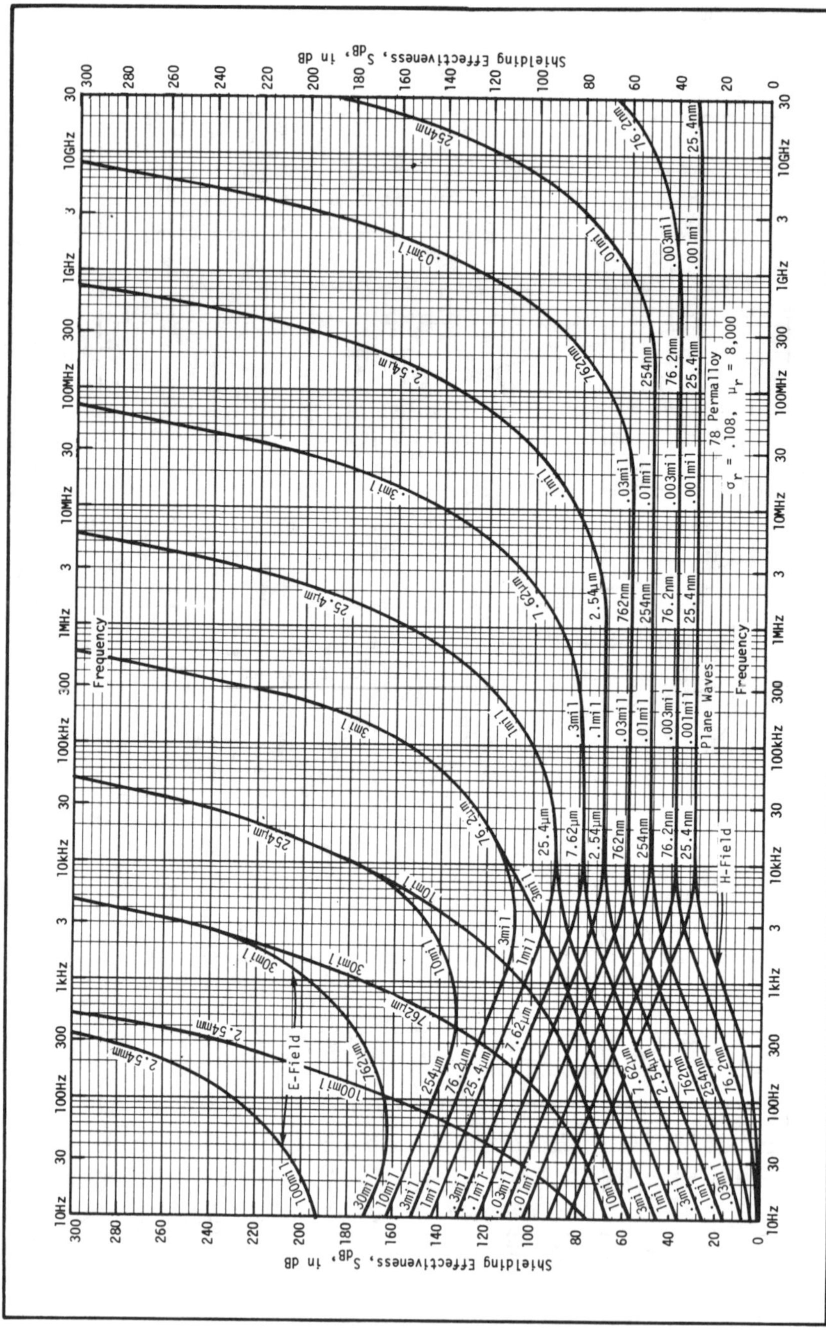

Figure F.6 - Shielding Effectiveness of 78 Permalloy vs. Frequency for Source-to-Metal Distance of 10km

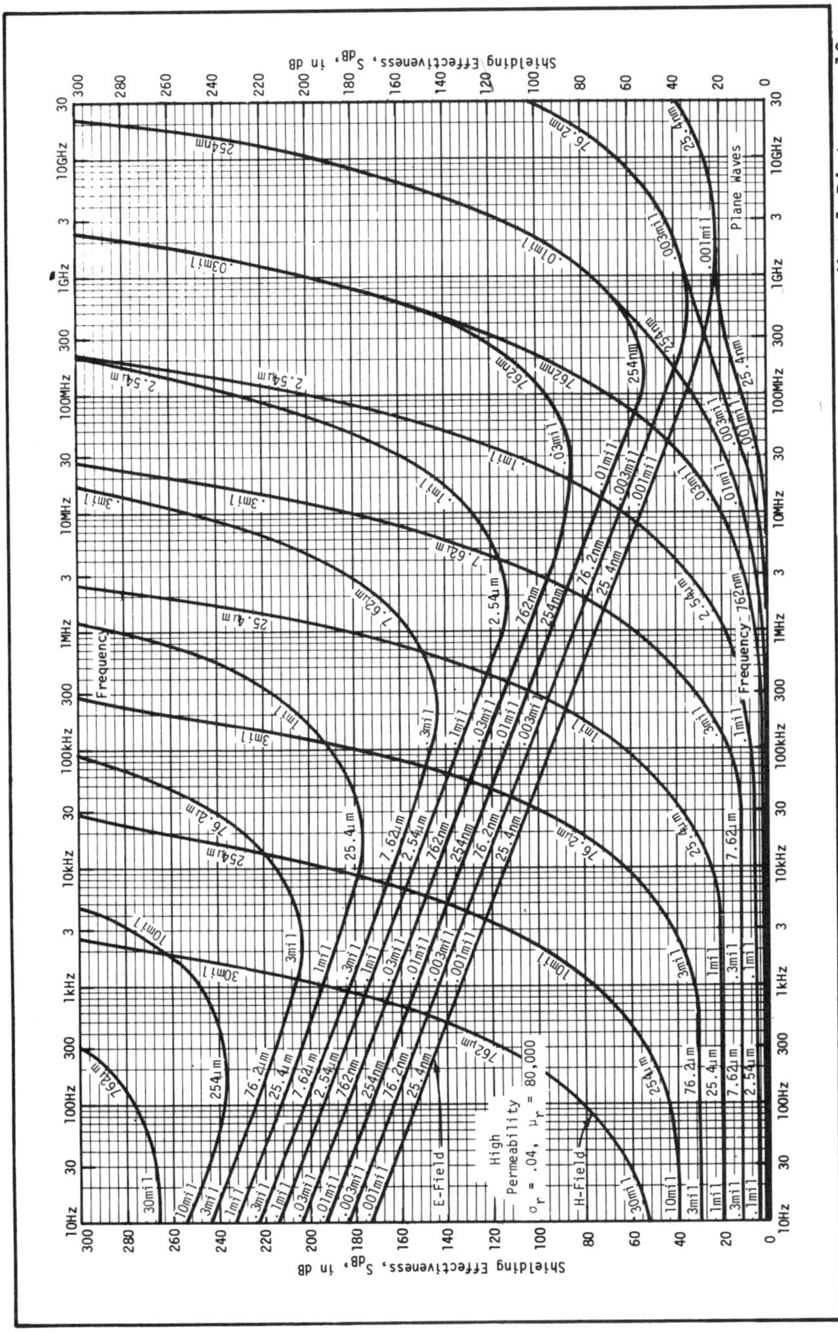

Figure G.1 - Shielding Effectiveness of High Permeability vs. Freq. for Source-to-Metal Distance: 10cm

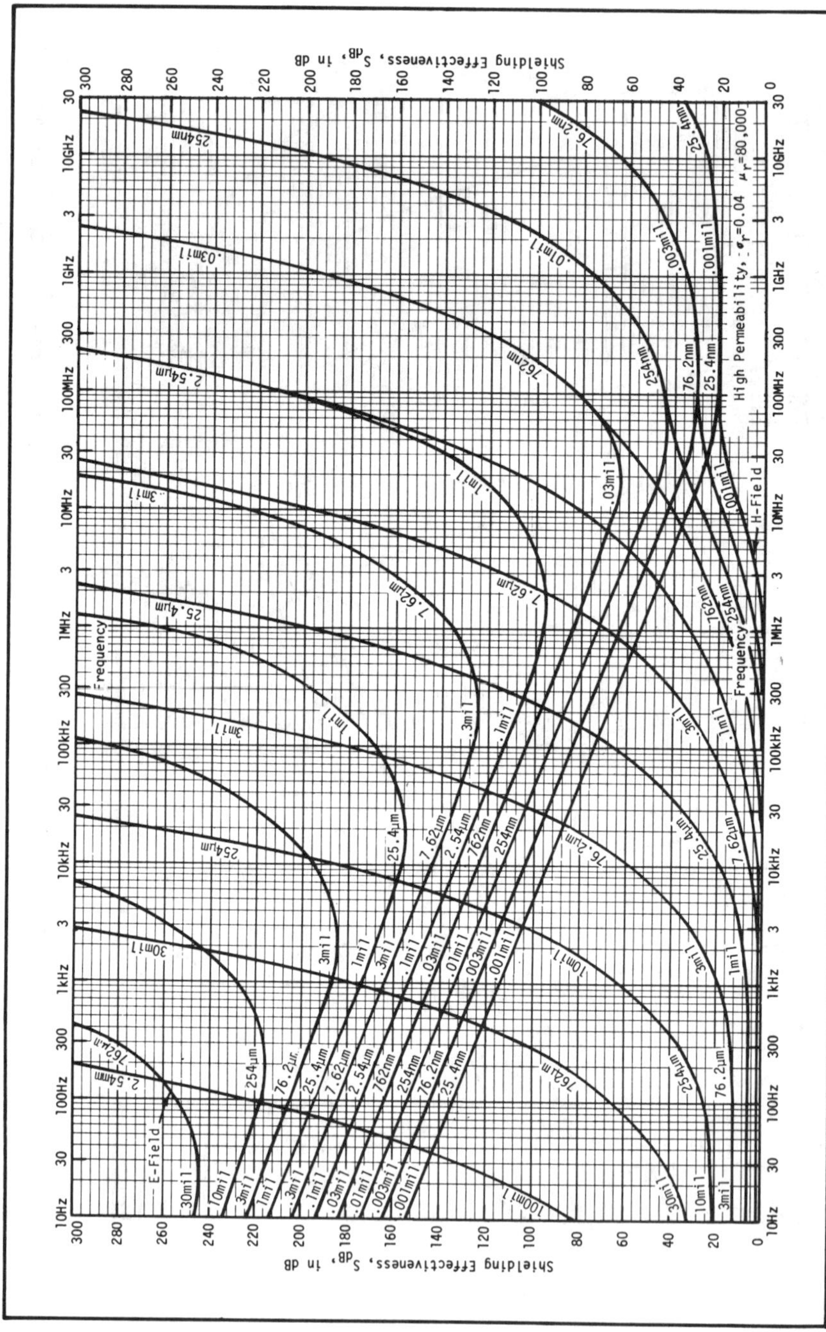

Figure G.2 - Shielding Effectiveness of High Permeability vs. Freq. for Source-to-Metal Distance of 1m

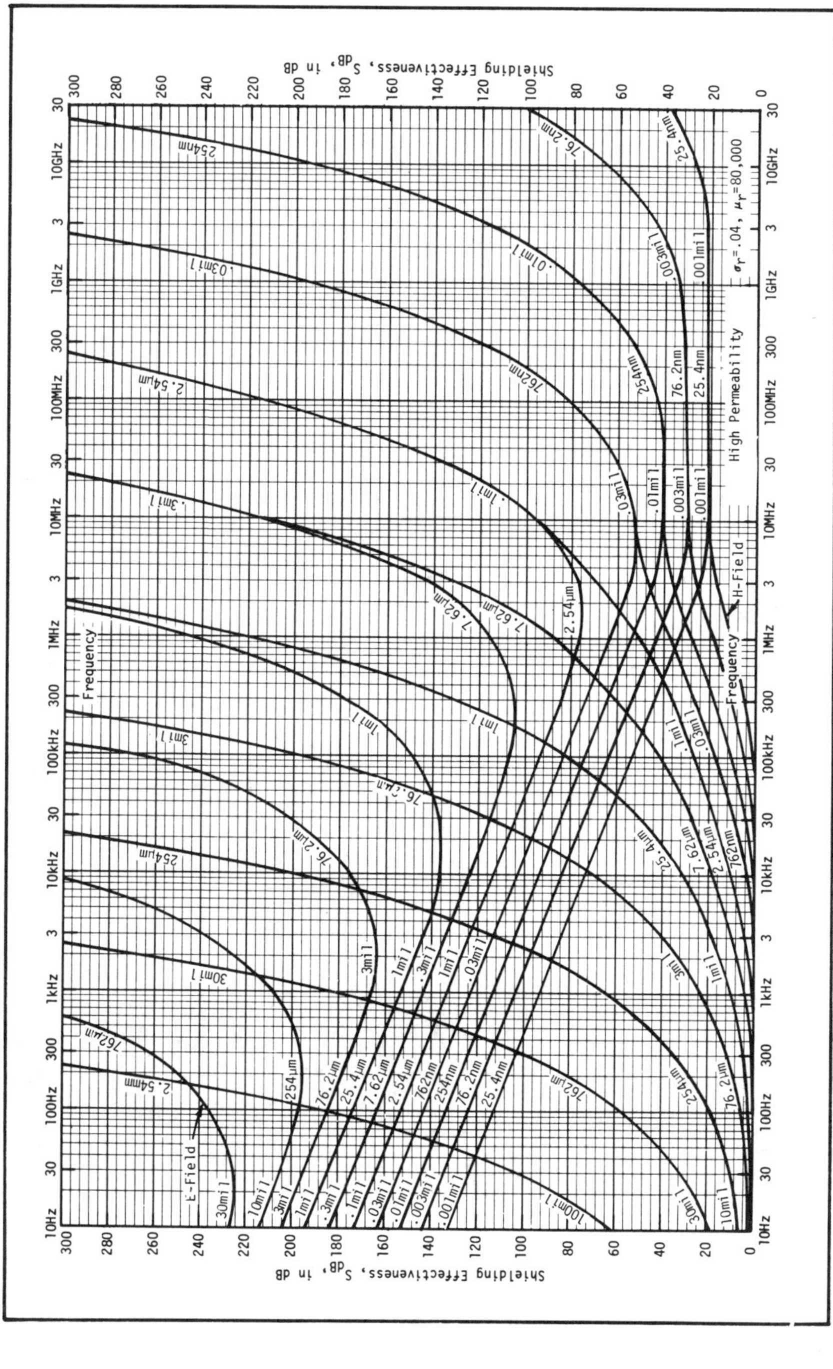

Figure G.3 - Shielding Effectiveness of High Permeability vs. Freq. for Source-to-Metal Distance: 10m

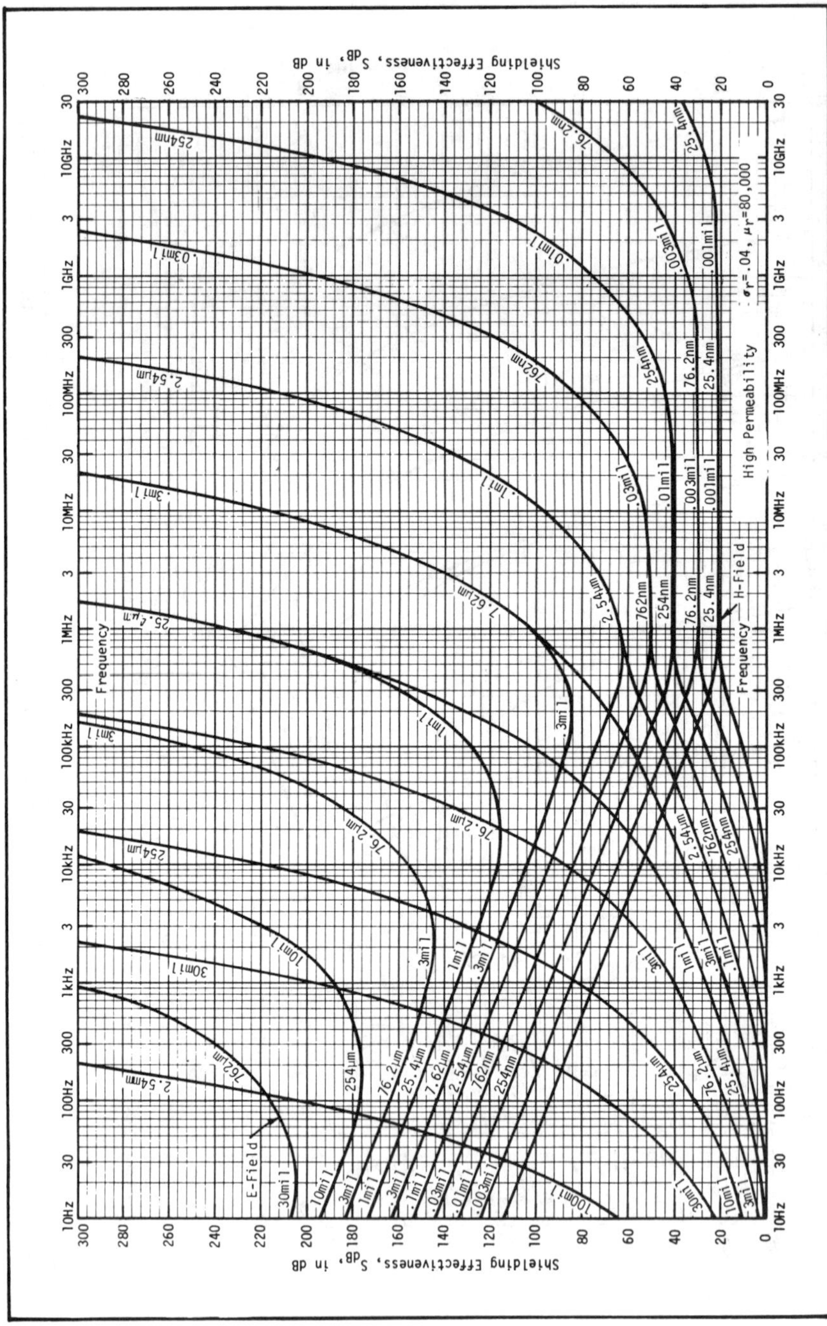

Figure G.4 - Shielding Effectiveness of High Permeability vs. Freq. for Source-to-Metal Distance: 100m

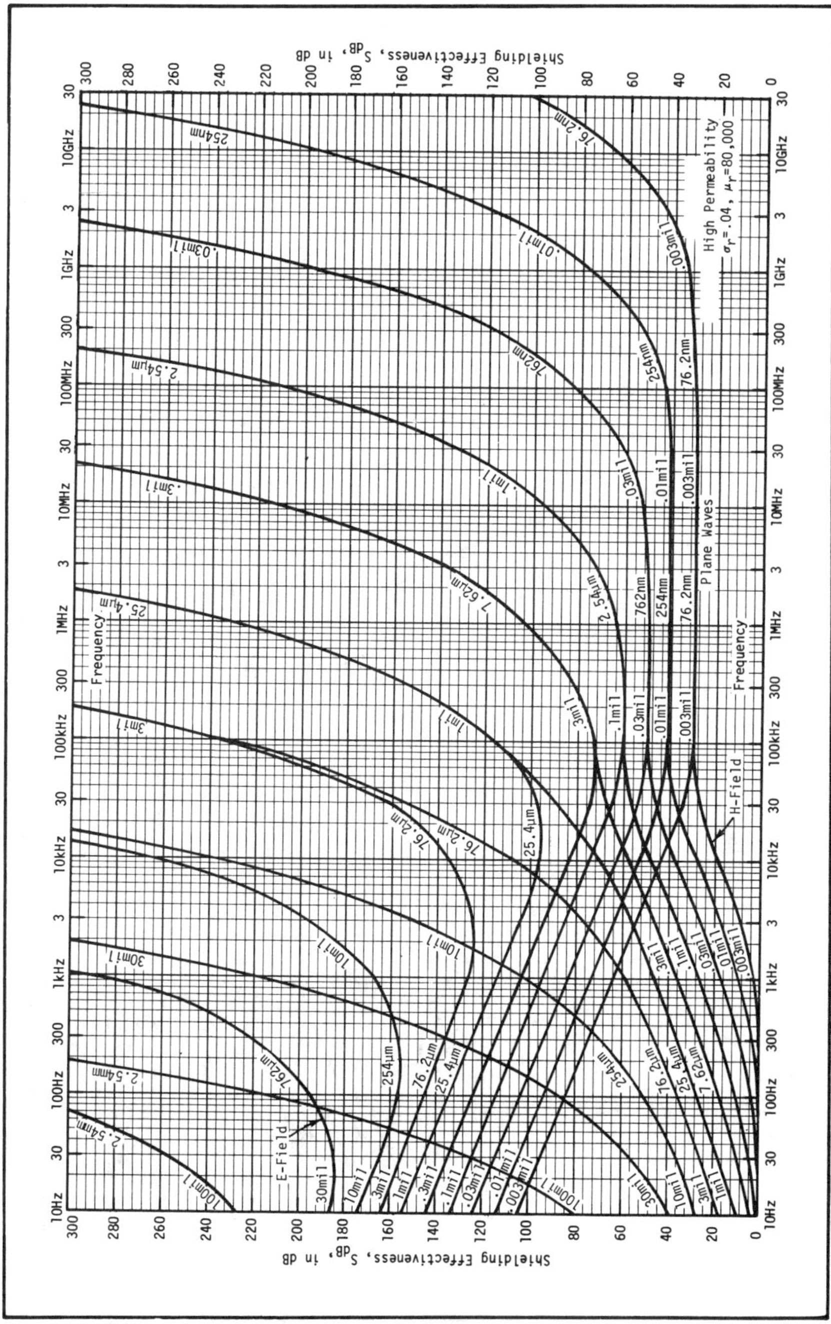

Figure G.5 - Shielding Effectiveness of High Permeability vs. Freq. for Source-to-Metal Distance: 1km

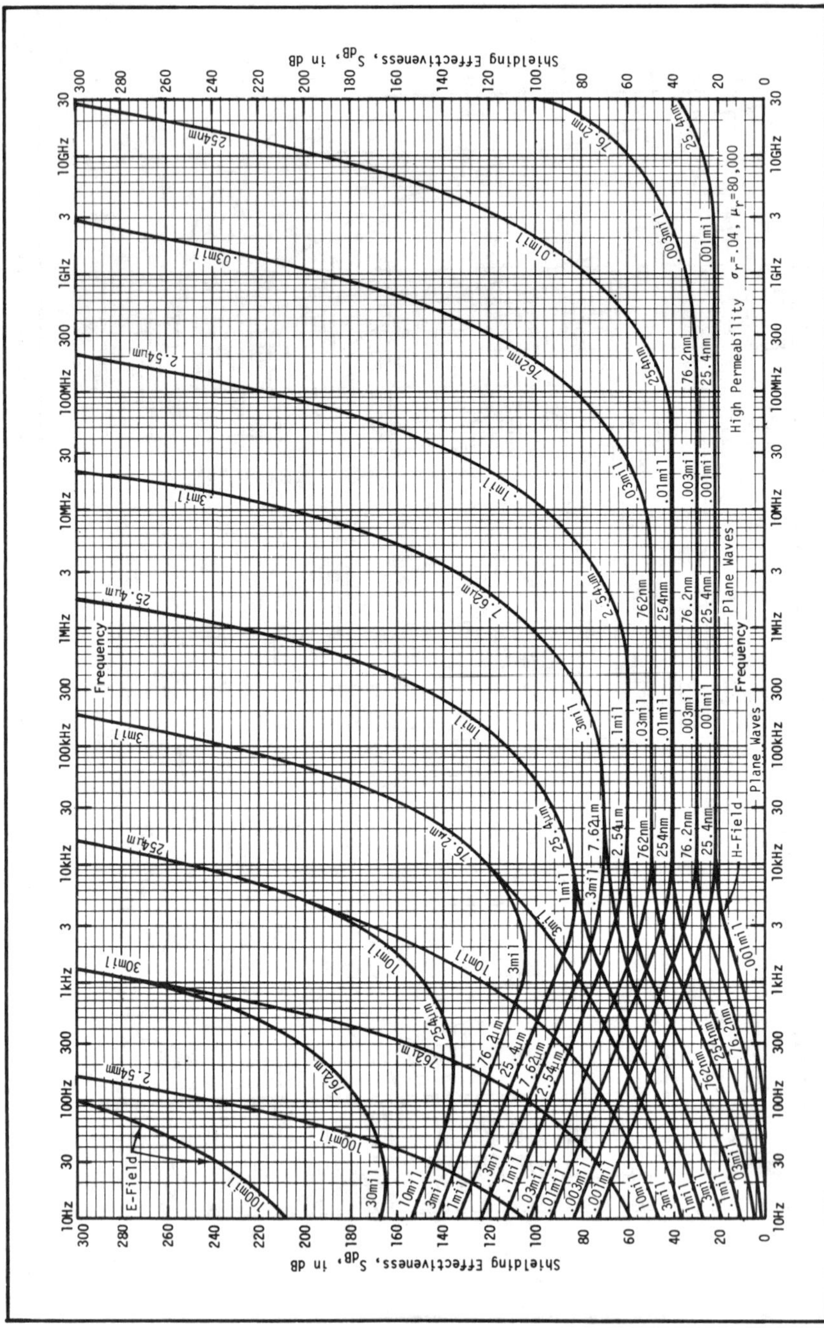

Figure G.6 - Shielding Effectiveness of High Permeability vs. Freq. for Source-to-Metal Distance: 10km

INDEX

Page No.

A

B

C

C (CONT'D)

D

E

F

F (CONT'D)

G

H

Index

I

K

L

M

M (CONT'D)

N

O

P

P (CONT'D)

R

S

S (CONT'D)

T

U-V

W

Z